Recent Titles from Quorum Books

Modern Analytical Auditing: Practical Guidance for Auditors and Accountants
Thomas E. McKee

Multiple Use Job Descriptions: A Guide to Analysis, Preparation, and Applications for Human Resource Managers
Philip C. Grant

Cost-Effective Marketing Research: A Guide for Marketing Managers
Eric J. Soares

Corporate Philosophies and Mission Statements: A Survey and Guide for Corporate Communicators and Management
Thomas A. Falsey

Strategic Executive Decisions: An Analysis of the Difference Between Theory and Practice
Michael J. Stahl

Motivating Strategies for Performance and Productivity: A Guide to Human Resource Development
Paul J. Champagne and R. Bruce McAfee

Innovation Through Technical and Scientific Information: Government and Industry Cooperation
Science and Public Policy Program, University of Oklahoma: Steven Ballard, Thomas E. James, Jr., Timothy I. Adams, Michael D. Devine, Lani L. Malysa, and Mark Meo

Transformations in French Business: Political, Economic, and Cultural Changes from 1981 to 1987
Judith Frommer and Janice McCormick, editors

Economics and Antitrust Policy
Robert J. Larner and James W. Meehan, Jr., editors

A Legal Guide to EDP Management
Michael C. Gemignani

U.S. Commercial Opportunities in the Soviet Union: Marketing, Production, and Strategic Planning Perspectives
Chris C. Carvounis and Brinda Z. Carvounis

An Analysis of the New Financial Institutions: Changing Technologies, Financial Structures, Distribution Systems, and Deregulation
Alan Gart

THE COMMERCIALIZATION OF OUTER SPACE

Opportunities and Obstacles for American Business

Jonathan N. Goodrich

QUORUM BOOKS
New York
Westport, Connecticut
London

Library of Congress Cataloging-in-Publication Data

Goodrich, Jonathan N.
The commercialization of outer space : opportunities and obstacles for American business / Jonathan N. Goodrich.
p. cm.
Bibliography: p.
Includes index.
ISBN 0–89930–342–0 (lib. bdg. : alk. paper)
1. Space commercialization—United States. I. Title.
HD9711.75.U62G66 1989
338.0919—dc19 88–23665

British Library Cataloguing in Publication Data is available.

Library of Congress Catalog Card Number: 88–23665
ISBN: 0–89930–342–0

First published in 1989 by Quorum Books

Greenwood Press, Inc.
88 Post Road West, Westport, Connecticut 06881

Printed in the United States of America

The paper used in this book complies with the Permanent Paper Standard issued by the National Information Standards Organization (Z39.48–1984).

10 9 8 7 6 5 4 3 2 1

To my parents, Rupert and Naomi Goodrich;
my wife, Grace; and my sons, Jason and Paul

Contents

Tables and Figures

TABLES

FIGURES

Abbreviations

Many of the following abbreviations appear in the text. Others are part of the space commercialization jargon.

AAAS	American Association for the Advancement of Science
AACB	Aeronautics and Astronautics Coordinating Board
ABM	Antiballistic Missile
ACC	Army Communications Command
ACSI	Assistant Chief of Staff for Intelligence
AFGL	Air Force Geophysics Laboratory
AFRPL	Air Force Rocket Propulsion Lab
AFSATCOM	Air Force Satellite Communications
AFSC	Air Force Space Command
AFSCF	Air Force Satellite Control Facility
AFWL	Air Force Weapons Lab
AIAA	American Institute of Aeronautics and Astronautics
ALCOA	Aluminum Corporation of America
ALMHV	Air-Launched Miniature Homing Vehicle
ANIK	Canadian Communications Satellite
ARABSAT	Arab Satellite Organization
ARPA	Advanced Research Projects Agency (DOD)

ASA (RD&A)	Assistant Secretary of the Army (Research, Development, and Acquisition)
ASAT	Antisatellite
ASAT MCC	Antisatellite Mission Control Center
ASPO	Army Space Program Office
ATM	*Apollo* Telescope Mount
AT&T	American Telephone & Telegraph
AUDRI	Automated Drug Identification (automated system for analyzing drugs and biological compounds)
BMD	Ballistic Missile Defense
BMDPM	Ballistic Missile Defense Program Manager
BMDPMO	Ballistic Missile Defense Program Manager's Office
BMDPO	Ballistic Missile Defense Program Office
BMEWS	Ballistic Missile Early Warning System
CBD	*Commerce Business Daily* (official announcement medium for federal procurements)
CBT SUP SYS	Combat Support System
CCC	Composite Consultation Concepts, Inc. (Houston, Texas)
C^3I	Command, Control, Communications, and Intelligence
C^4	Command, Control, Communications, and Computers
CHM	Common Heritage of Mankind ("Moon Treaty")
CINCNORAD/ ADCOM	Commander in Chief, North American Aerospace Defense Command
CNRET	U.N. Centre for Natural Resources, Energy, and Transport
COMM POS/NAV	Communication Position/Navigation
COMSAT	Communications Satellite Corporation
COMSTAR	Communications Satellite (major contractor, Hughes Aircraft)
COPUOS	UN Committee on Peaceful Uses of Outer Space
COSMIC	Computer Software Management and Information Center, University of Georgia
CPFF	Cost-Plus-Fixed-Fee
CSAF	Chief of Staff, U.S. Air Force
CSC	U.S. Civil Service Commission
CSOC	Consolidated Space Operations Center
DARPA	Defense Advance Research Projects Agency
DCSOPS	Deputy Chief of Staff for Operations and Plans

DCSRDA	Deputy Chief of Staff for Research, Development and Acquisition
DDN	Defense Digital Network
DEFCON	Defense Condition
DFRF	Dryden Flight Research Facility (NASA)
DMSP	Defense Meteorological Satellite Program
DOD	U.S. Department of Defense
DOT	U.S. Department of Transportation
DSAT	Defense Satellite
DSCS	Defense Satellite Communications System
DSN	Deep Space Network
EBU	European Broadcasting Union
EEC	European Economic Community
ELDO	European Launcher Development Organization
ELINT	Electronic Intelligence
ELV	Expendable Launch Vehicle
EORSAT	Electronic Warfare Data Relay Satellite (Electro-optical Rectifier)
ERL	Earth Resources Laboratory (NASA)
ESA	European Space Agency
ESD	Electronic Systems Division
ESMC	Eastern Space/Missile Center
ESRO	European Space Research Organization
EURECA	European Retrievable Carrier (space platform)
EUROS	European Automated Platform (space platform)
FAA	Federal Aviation Agency (or Administration)
FCC	Federal Communications Commission
FDA	U.S. Food and Drug Administration
FD&C Act	Food, Drug and Cosmetic Act
FEL	Free-Electron Laser
FEP	Fluorinated Ethylene Propylene
	Front End Processor
	Fluoral Ethel Propane
FLTSATCOM	Fleet Satellite Communications System
FRG	Federal Republic of Germany
FSEC	Florida Solar Energy Center (Dink Company, Alachua, Florida)
FTC	U.S. Federal Trade Commission
GAO	U.S. General Accounting Office

GAPSAT	Geological Applications Program Satellite
GE	General Electric Corporation
GEO	Geostationary Orbit
GLONASS	Soviet Satellite Navigation System
GMPs	Good Manufacturing Practices
GOES	Geostationary Operational Environmental Satellite
GPS	Global Positioning System
GPSCS	General Purpose Satellite Communications System
GSA	General Services Administration
GSFC	Goddard Space Flight Center (NASA, Greenbelt, Maryland)
HOE	Homing Overlay Experiment
IAA	International Academy of Astronautics
IACs	Industrial Applications Centers (NASA)
IAEA	International Atomic Energy Agency
IAF	International Astronautical Federation
IBM	International Business Machines
ICAO	International Civil Aviation Organization
ICBM	Intercontinental Ballistic Missile
ICSU	International Council of Scientific Unions
IFB	Invitation for Bids
IISL	International Institute of Space Law
ILO	International Labor Organization
IMCO	Inter-Governmental Maritime Consultative Organization
INCA	Intelligence Communications Architecture
INS	Inertial Navigation System
INSCOM	Intelligence and Security Command
INTEC	International Technology Underwriters (a leading U.S. space insurance firm)
INTELSAT	International Telecommunications Satellite Organization (provides satellite communications between nations)
INTERSPUTNIK	International System and Organization of Space Communications
ISF	Industrial Space Facility (space platform built by Space Industries, Inc., Houston, Texas)
ITU	International Telecommunications Union

JEA	Joint Endeavor Agreement (between NASA and private contractors)
JPL	Jet Propulsion Laboratory (NASA, Pasadena, California)
JSC	Johnson Space Center (NASA, Houston, Texas)
KSC	Kennedy Space Center (NASA, Cape Canaveral, Florida)
LDC	Less Developed Country
LEASAT	Leased Satellite
LEO	Low Earth Orbit
LOR	Lunar Orbit Rendezvous
LORAN	Long-Range Navigation
MARISAT	Maritime Satellite
MBB	Messerschmitt-Bolkow-Blohm (West German company that makes space platforms and other space hardware)
MCC	Mission Control Center
MDAC	McDonnell Douglas Corporation
MILSATCOM	Military Satellite Communications
MILSTAR	Extremely High Frequency Communications Satellite System
Mir ("*Peace*")	Name of a USSR Space Station
3M	Minnesota Mining and Manufacturing
MMPS	Manufacturing and Materials Processing in Space
MOL	Manned Orbiting Laboratory (Department of Defense)
MOU	Memorandum/Memoranda of Understanding (between NASA and private contractors)
MPS	Materials Processing in Space
MSC	Manned Spacecraft Center
MSF	Manned Space Flight
MSFC	Marshall Space Flight Center (NASA, Huntsville, Alabama)
MTF	Man-tended Flyer (space platform)
NAG	Naval Astronautics Group
NAS Act	National Aeronautics and Space Act (1958)
NASA	National Aeronautics and Space Administration
NASDA	National Space Development Agency (Japan)
NATO	North Atlantic Treaty Organization

NAVELECS	Naval Electronics System Command
NAVSPACECOM	Naval Space Command
NAVSPASUR	Naval Space Surveillance System
NAVSPASYSACT	Naval Space Systems Activity
NAVSTAR	Navigation and Traffic Control System Satellite
NERAC	New England Research Applications Center (NASA center, Storrs, Connecticut)
NERVA	Nuclear Engine for Rocket Vehicle Application
NOAA	National Oceanographic and Atmospheric Administration
NORAD/ADCOM	North American Air Defense Command/Aerospace Defense Command
NSF	National Science Foundation
NSTL	National Space Technology Laboratories (NASA, NSTL Station, Mississippi 39529)
NTIS	National Technical Information Service
OACSI	Office of the Assistant Chief of Staff for Intelligence
OAO	Orbiting Astronomical Observatory (NASA)
OART	Office of Advanced Research and Technology (NASA)
OCST	Office of Commercial Space Transportation
OMB	Office of Management and Budget
OMSF	Office of Manned Space Flight (NASA)
OMV	Orbital Maneuvering Vehicle
OSHA	Occupational Safety and Health Administration
OSSA	Office of Space Science and Applications (NASA)
OTA	Office of Technology Assessment
OTDA	Office of Tracking and Data Acquisitions (NASA)
PAM	Payload Assist Module
PBI	Polybenzimidazole (fireblocking fiber)
PCG	Planning Coordination Group (NASA)
PCSG	Planning Coordination Steering Group (NASA)
PERT	Program Evaluation and Review Technique
PFP	Program and Financial Plan
POS/NAV	Communication Position/Navigation
P^3I	Preplanned Product Improvements
PPBS	Planning-Progamming-Budgeting System
PPP	Phased Project Planning

PPS	Precise Positioning Service
PRC	People's Republic of China
PRESTO	Performance of Regenerative Superheated Steam Turbine Cycles (computer software package produced by COSMIC)
PSAC	President's Science Advisory Committee
PSG	Planning Steering Group (NASA)
RCA	Radio Corporation of America
R&D	Research & Development
RDLP	Research and Development Limited Partnership
RDS	Director of Space Systems and Command, Control, and Communications
RFP	Request for Proposal
RIF	Reduction in Force
RMS	Remote Manipulator System
ROI	Return on Investment
RORSAT	Radar Ocean Reconnaissance Satellite
SAFUS	Under Secretary of the Air Force
SALT 1	1972 Arms Control Agreement between the United States and the USSR
SATCOM	Satellite Communications (part of Defense Satellite Communications System)
SATKA	Surveillance Acquisition, Tracking, and Kill Assessment
SBA	Small Business Administration
SBKKVs	Space-based Kinetic Kill Vehicles
SBS	Satellite Business Systems, Inc.
SCF	Satellite Control Facility
SCORE	Signal Communications by Orbiting Relay Equipment
SD	Space Division
SDI	Strategic Defense Initiative
SDIO	Strategic Defense Initiative Office/Organization
SDS	Shuttle Dynamic Simulation (Simulator)
	Software Design Specifications
	Steering Damping System
SEB	Source Evaluation Board
SEC	Security and Exchange Commission
SEWS	Satellite Early Warning System

SHF	Super High Frequency
SIG	Senior Interagency Group
SLBM	Submarine-Launched Ballistic Missile
SMF	Space Manufacturing Facility
SPADATS	Space Detection and Tracking System
SPADOC	Space Defense Operations Center (Cheyenne Mountain, Colorado Springs, Colorado)
SPAR	Structural Performance and Design (computer program developed by NASA's Lewis Research Center, Cleveland, Ohio)
SPAS	Shuttle Pallet Satellites (space platforms)
SPOT	French Remote Sensing Satellite
SPS	Solar Power Satellite
SPS	Standard Positioning Signal
SRB	Solid Rocket Booster
SRT	Supporting Research and Technology
SSASC	Space Science and Applications Steering Committee (NASA)
SSI	Space Services Inc.
SSS	Sound Suppression System
	Space Shuttle System
	Stage Separation Subsystem
	Station Set Specification
	Subsystem Segment
SST	Supersonic Transport
ST	Space Telescope
STAC	Science & Technology Advisory Committee (NASA)
STG	(President's) Space Task Group
STS	Space Transportation System
TACAN	Tactical Air Command and Navigation System
	Tactical Air Navigation
TDRSS	Tracking and Data Relay Satellite System
TEA	Technical Exchange Agreements
TENCAP	Tactical Exploitation of Naval Space Capabilities
TIE	Technical Integration and Evaluation
TOS	Transfer Orbit Stage
TRADOC	Training and Doctrine Command
TWA	Trans-World Airlines

UFO	Unidentified Flying Object
UK	United Kingdom
UN	United Nations
UNDP	United Nations Development Programme
UNDRO	UN Disaster Relief Organization
UNEP	UN Environment Programme
UNESCO	United Nations Educational, Scientific, and Cultural Organization
Unispace '82	Second United Nations Conference on the Exploration and Peaceful Uses of Outer Space
UNITAR	UN Institute for Training and Research
USAU	U.S. Aviation Underwriters
USDA	U.S. Department of Agriculture
VAFB	Vandenberg Air Force Base
VCSA	Vice-Chief of Staff, U.S. Army
VOR/DME	Very High Frequency Omni-directional Range Distance-Measuring Equipment
V/STOL	Vertical or Short Take-off and Landing (aircraft)
WESTAR	Western Union Satellite
WIPO	World Intellectual Property Organization
WMO	World Meteorological Organization
WSMC	Western Space/Missile Center
WWDSA	Worldwide Digital System Architecture
XOS	Directorate of Space Operations

Preface

Space commercialization is important to the United States and is of growing interest to other nations, such as the USSR, China, Japan, the UK, Canada, and Brazil. This book, unlike others on space, deals primarily with marketing, economic, legal, insurance, and other business issues of concern to American executives in the space industry and to those who contemplate entering the arena of space commercialization.

The book should be useful to marketing executives in particular, and business and government executives in general, in the space industry. The industry includes space agencies, such as the National Aeronautics and Space Administration (NASA); aerospace firms, such as McDonnell Douglas, Lockheed Corporation, and the Boeing Company; and nonaerospace firms that have a vested interest in using the space environment for materials processing in space (MPS) and/or for producing products to be used in space, such as IBM (computers) and RCA (satellites).

Chapter 1, "The Uses and Importance of Outer Space to the United States," outlines the basic goals of U.S. space policy and discusses the military and civilian uses of outer space. The chapter also mentions future uses of outer space.

Chapter 2, "Economic Opportunities in and Obstacles to Space Commercialization: An Overview," is a profile of space commercialization. It discusses major components of space commercialization or business opportunities in space commercialization (e.g., transportation and launch services, satellite communications, and defense), obstacles to space business, and marketing implications of the opportunities and obstacles. The chapter lays the foundation for the rest of the book. Topics mentioned briefly in the chapter are dealt with more exhaustively in separate chapters.

Chapter 3, "Space Operators," provides an overview of the principal operators in space—private enterprise, government agencies, and international agencies—and their working relationships.

Chapter 4, "Materials Processing in Space," discusses potential MPS products and markets, and obstacles to MPS. Chapter 5, "Spin-offs and Market Opportunities," discusses some of the major spin-offs of space technology that have found their way into the marketplace and affect our lives.

Chapter 6, "SDI Is Big Business," looks at military components of the SDI and provides insights into how it is big business in the United States and other countries. Chapter 7, "A Methodology for Identifying Customers for In-Orbit Facilities," identifies methods that private sector and public sector makers of space stations and space platforms can use to identify industrial customers for these in-orbit space facilities. This is a crucial area of space marketing.

Chapter 8, "How to Market to NASA," is aimed at acquainting organizations (prospective contractors) with how to market their products and services to NASA, whether it be an abstract idea, a manufacturing capability, a fabricated component, construction, basic materials, or a specialized service. It deals with the nature of marketing; knowing the market (NASA); NASA's field installations, needs, and procurement process; and the importance of marketing your organization's capabilities.

Chapter 9 discusses types of space insurance and the role and importance of space insurance in space commercialization. Chapter 10 discusses major space laws and their implications for space commerce.

The Conclusion looks to provocative areas of space commerce, among them the search for extraterrestrial intelligence, Third World activities in space, the prospects for Soviet and U.S. cooperation in space ventures, and the next fifty years.

Most of the chapters provide marketing insights into space commercialization, especially chapters 2, 4, 5, 7 and 8.

The book contains other sections useful to readers: a list of abbreviations of key terms used in the space commerce literature; appendixes identifying space business publications and firms involved in space commerce; a glossary; and a reference list of some 400 sources on space commerce.

Acknowledgments

I owe a debt of gratitude to many colleagues and friends who reviewed drafts of the manuscript and provided helpful comments and suggestions that strengthened the book. I would especially like to thank Gary H. Kitmacher, aerospace flight systems engineer at NASA/Johnson Space Center, for introducing me to the subject of space commerce a few years ago and for being a valuable source of facts and ideas. Dennis J. Gayle, Florida International University, provided helpful comments about the manuscript and contributed to the writing of chapter 7; and Jeff Hetherington contributed to the writing of chapter 9. Lisa M. Wong helped in the research. Many people at various NASA locations also provided space documents and insightful ideas.

I wish to thank many typists who participated in the preparation of the final manuscript: Irene Young, Toby Levin, Tere Pastoriza, Hilda Aguiar, and Yasmine Paggi. I would also like to thank Thomas Gannon, acquisitions editor of Quorum Books, for his suggestion that I write this book and for his encouragement.

Finally, I owe a special debt of thanks to my wife, Grace, and our two children, Jason and Paul. Their understanding, cooperation, and support during the writing of this book are sincerely appreciated.

Introduction

This is *not* another book about the courage, daring, and life of astronauts, such as Tom Wolfe's *The Right Stuff* (1979). It is *not* a personal account of one astronaut's journey into space, such as *Gemini: A Personal Account of Man's Venture into Space* (1968), by the late Virgil "Gus" Grissom. These books, and others like them, have their place in history. *The Commercialization of Outer Space: Opportunities and Obstacles for American Business* is a book that shows the marketing executive in particular, and the business executive in general, the economic realities, opportunities, and obstacles of space business.

The book is a back-to-basics approach to business decision making and strategy for executives involved in space business and those contemplating getting involved. Rather than being fascinated by the glamour and publicity of space business, such executives must ask basic questions that affect the bottom line. Is there a market in space for our product? If so, how large is it? Does the market size make the space venture economically profitable? Will Earthbound production techniques do just as well as outer space production techniques? Will space-made products be too expensive for Earth or space markets? Is space transportation reliable, with on-time schedules? Are space transportation costs too high to justify production in space? What are the revenue streams? Is the payback period too long and risky? Can American companies rely on the federal government's space policies, which can change with new U.S. presidents? Is space insurance available? from whom? at what cost? Will space law be a deterrent to doing business in outer space? What about product liability and proprietary rights of space-made products or products made on Earth for use in space? These are just a few of the questions executives in the space business will have to address prior to getting involved in space ventures.

This book deals with these issues; there are chapters on such varied topics as materials processing in space (MPS), space insurance, space law, spin-offs, and how to market to the National Aeronautics and Space Administration (NASA). It is the first book of its kind to address, in one volume, these many relevant business aspects of the commercialization of outer space. To this end, I hope that the book will be instructive, useful, and provocative reading for executives in the aerospace and related industries, companies that market goods and services to NASA, and companies that market spin-off products from space technology to consumer and industrial markets, and companies contemplating space commercialization ventures.

THE COMMERCIALIZATION OF OUTER SPACE

CHAPTER 1

The Uses and Importance of Outer Space to the United States

People have always been fascinated by the unknown and the challenge of outer space. In ancient times, they talked about outer space in terms of the stars and the heavens. Today we have more precise knowledge about outer space but still very little in terms of the larger scheme of things.

Outer space begins where the Earth's atmosphere is too thin to affect objects moving through it. Near the Earth's surface, air is plentiful, but higher above the Earth, the air becomes thinner and thinner. Little by little, the air fades to almost nothing, and space begins. Scientists generally regard outer space as beginning around 50 miles above the Earth. Even there, however, there is still enough air present to slow a satellite and eventually cause it to fall. In addition, solar storms in the upper atmosphere may cause satellites to fall sooner than expected. Outer space consists of the space between planets, stars, moons, and galaxies, as well as those bodies themselves.

This chapter provides an insight into the uses and importance of outer space to the United States. Other world powers, such as the USSR, Japan, Europe, and the People's Republic of China (PRC) have similar uses for outer space that will be described here, but the emphases may differ.

BASIC GOALS OF U.S. SPACE POLICY

The basic goals of U.S. space policy are to:

1. Strengthen the security of the United States.
2. Maintain U.S. space leadership.

3. Obtain economic and scientific benefits through the exploitation of space.
4. Expand U.S. private sector investment and involvement in civil space and space-related activities.
5. Promote international cooperative activities in the national interest.
6. Cooperate with other nations in maintaining the freedom of space for activities that increase the security and welfare of humanity.
7. Establish human and robotic settlements in space (e.g., on the Moon and Mars) during the twenty-first century and beyond.

Many principles underlie the conduct of the U.S. space program. The major principles are:

1. The United States is committed to the exploration and use of space by all nations for peaceful purposes and for the benefit of humanity.
2. The United States rejects any claims of sovereignty by any nation over space or over celestial bodies, or any portion thereof, and rejects any limitation on the fundamental right to acquire data from space.
3. The United States encourages the commercialization of space, consistent with national security concerns, treaties, and international agreements.
4. The U.S. space program consists of two distinct programs: military (national security) and civil.
5. The United States will pursue its activities in space in support of its right to self-defense.
6. The United States will continue to study space arms control.

USES OF OUTER SPACE

Military Uses

The United States has two separate space programs, one civilian and the other military, one open to the world's scrutiny and the other closed. The civilian space program encompasses such efforts as launching weather satellites to help forecast weather conditions, encouraging American businesses to conduct scientific and industrial experiments in space, and exploring the Solar System and the universe to advance scientific knowledge. The military space program seeks to exploit and command the "new high ground" (Karas 1983). This offers the United States three strategic assets: greater tactical strength, protection from access, and a wider view of Earth and space. The major impetus for the military space program is to increase America's military strength and protect it from potential aggression by the Soviet Union.

Space is vital to the conduct of U.S. military operations because weapons of destruction can be deployed from space, military control of Earth may be possible someday from outer space, and because the information and data necessary to verify treaties, conduct a modern war, and so on depend to an increasing degree on satellites in space. The United States can be so prepared only by its presence and dominance in outer space.

Military uses of space include the following (Downey 1985, p. 27):

- Surveillance
- Attack warning and assessment
- Communication
- Navigation
- Meteorology
- Geodesy
- Space defense

The United States has space systems that already perform these missions. These systems are updated continually with sophisticated technology that increases accuracy and power. Let's examine briefly each of these military uses of space.

Surveillance. This refers to regular monitoring activity. In the 1950s, prior to the era of satellites, the Lockheed U–2 reconnaissance aircraft was created in an extraordinary technical tour de force by Kelly Johnson and his collaborators. For a number of years, these remarkable airplanes flew over denied territory with impunity because they could operate at such extremely high altitudes that antiaircraft fire could not reach them. In 1960, however, the U-2 era was brought to a close by the Soviets' downing of a U-2 airplane flown by Francis Gary Powers.

At about that time, Earth-orbiting satellites that could perform similar surveillance functions were almost ready to be deployed. Since the first of these was launched in the early 1960s, they have played an increasingly important role. Much information concerning these surveillance systems in space is classified. Generally, however, surveillance satellites—which usually weigh anywhere from two to five tons and cost from $50 million to $250 million each—can photograph enemy ground forces and warships, warn against surprise missile attacks, and detect sneak testing of nuclear bombs in space by other nations. In 1972, the United States and the Soviet Union agreed in the Arms Control Agreement (SALT 1) not to attack each other's "national technical means of verification," the euphemism then employed for photoreconnaissance satellites.

The United States and the USSR have promised not to place offensive military weapons in space, but these promises are as brittle as politicians' oaths because neither side trusts the other. Additionally each side is always seeking the military edge—as well each should. Both countries have developed antisatellite systems that can interfere with the other's satellites and other nations' satellites, and both are working on a space-based missile defense system, the Strategic Defense Initiative (SDI), dubbed "Star Wars."

Attack Warning and Assessment. The United States employs satellites to provide early warning of a nuclear attack. These satellites, in geosynchronous equatorial orbit (22,300 miles altitude above the equator), detect missiles in the boost stage of flight using infrared sensors. Major efforts are underway to make the system as robust and reliable as possible. For example, mobile, survivable ground terminals are being developed to reduce system dependence on fixed ground stations, and surveillance satellites are being improved to reduce their vulnerability to missiles, laser attacks, and other countermeasures.

Communication. The United States has several communications and surveillance satellites in various orbits watching planet Earth. They are in polar and geosynchronous orbits and over the Atlantic, Pacific, and Indian oceans. They are used to transmit communications swiftly across international frontiers. With military forces deployed around the world, it is both technically efficient and politically advantageous to use satellite relays instead of terrestrial communications. Today about 80 percent of all military long-distance communications are transmitted via satellites—testimony to the importance of space for military command, control, and strategy.

Many of these satellites have acronyms and are designed for special use by the U.S. Navy, Army, and Air Force. Examples are the Defense Satellite Communications System (DSCS), the Air Force Satellite Communications System (AFSATCOM), the Fleet Satellite Communications System (FLTSATCOM), and a powerful new system called MILSTAR. MILSTAR provides coordinated, low-capacity, jam-resistant communications for U.S. strategic and tactical forces worldwide, encompassing all the U.S. armed forces except the Coast Guard. The Earth terminal segment of the MILSTAR system consists of terminals located on aircraft, ships, submarines, armored track vehicles, and jeeps and at fixed ground sites.

Navigation. Navigation is the fourth military use of space using satellites. Navigation satellites help pilots and sailors to find their exact location in all kinds of weather. The United States launched its first ocean navigation satellite in April 1960, the beginning of the TRANSIT series of satellites, whose principal purpose was to provide accurate location fixes for the inertial navigation systems of submarines carrying sea-launched ballistic missiles (SLBMs). In addition to TRANSIT satellites, the U.S. Navy has developed an improved, navigational space-based satellite system, NAVSTAR/GPS (Global Positioning System). The GPS space segment consists of eighteen satellites and hosts the nuclear detonation detection payloads.

Meteorology. The Defense Meteorological Satellite Program (DMSP) supports Department of Defense (DOD) needs for weather information. DMSP satellites in 500-mile altitude orbit provide global weather information from all points on the Earth to key locations worldwide to support U.S. Army, Navy, and Air Force tactical operations. Examples of meteorological data that help in defense operations are cloud cover data used to assist reconnaissance missions; more accurate weather forecasts to improve planning of military operations; and wind speed, direction, and precipitation to assist in forecasting the effects of smoke, gas, or nuclear weapons.

Geodesy. Space-based satellite systems provide photographic and other data about the Earth that are used to produce military maps. Accurate mapping is essential for such purposes as precise ballistic missile launch and impact point location and for strategic command purposes.

Space Defense and Antisatellite Systems. All the previously mentioned military uses of space—surveillance, attack warning and assessment, and navigation—contribute to this last military use of space: space defense and antisatellite systems. It is given separate billing because it has become, and will continue to be, of paramount importance in America's defense strategy. In fact, space defense and antisatellite systems are core components of what is now popularly known as SDI.

The objectives of space defense and antisatellite systems are to provide improved surveillance of space, including warning that U.S. space systems are being attacked; to increase the survivability of U.S. military satellites and spacecraft; to develop space defense weapons systems; and to develop an adequate command and control system for space defense. The program will require space-based defense outposts and space stations for observation purposes, as well as for deployment of weapons and artillery and Earth-based military installations.

The U.S. Air Force has established a Space Defense Operations Center (known as SPADOC) in Cheyenne Mountain, Colorado Springs, Colorado, to serve as the hub for surveillance information from the Space Detection and Tracking System (SPADATS), missile warning sensors, and information from other sources.

Space-based lasers and directed energy weapons are also being developed and investigated to determine their utility for space defense. The survivability of America's space systems is also being improved through proliferation of satellites with decoys; camouflage and deception; greater autonomy in the satellites; increased shielding against nuclear and electromagnetic radiation; and additional maneuver capability. None of these techniques, singly or in combination, can offer absolute safety, but they can act as a deterrent to an attacker or make life more miserable for that country(ies).

Civilian Uses of Space

Generally the broad civilian use of space is to exploit commercial opportunities in space to benefit humanity. For example, satellites are used to transmit radio and television broadcasts all over the world. We are expanding our knowledge of Earth and other planets through flybys, prospector missions, and astronomical studies of celestial bodies in space. Some of the major civilian uses of space are similar to military uses of space; they will be listed but not dealt with in any detail.

Communication. Communications satellites for firms, including Radio Corporation of America (RCA) and American Telephone and Telegraph Company (AT&T), are used to transmit radio and television programs, telephone conversations, and other data. Satellites owned by countries, such as Mexico, Australia, France, Indonesia, and the USSR, are used to transmit similar data (and military intelligence).

Earth Observation. A variety of satellites, such as remote-sensing satellites and weather satellites, are equipped with state-of-the-art cameras, radars, and special sensors to collect data about the atmosphere (for weather forecasting and airline flights), land (for forecasting earthquakes, volcanos, and other such phenomena), the seas (for maritime activities and fish movement), and other activities on Earth.

Advancing Science. One of the most challenging and promising uses of space, and a goal of U.S. space policy, is advancing scientific knowledge about Earth, other planets, the Solar System, and the universe (or universes). We have learned much about these through robotic excursions into space (space probes), flyby missions, astronomy studies, and human journeys into space on the U.S. space shuttles (*Atlantis*, *Discovery*, *Columbia*, *Challenger*), and on USSR space stations *Salyut 1* through 7, and *Mir* ("Peace").

An example of a space probe is *Voyager* 2, launched from Cape Canaveral, Florida, on August 20, 1977 (*Miami Herald* 1987d). It sends back data about planets and the Solar System. *Voyager* 2, zipping along at 43,000 mph, will pass within 3,100 miles of Neptune and within 25,000 miles of that planet's large moon, Triton, in August 1989. It will continue out of the Solar System, leaving only Pluto unvisited by probes.

NASA is also planning a robot spacecraft for a fifty-year mission into unexplored deep space beyond the Solar System. TAU, as the project is called, needs ion rockets that will fire continuously for ten years, slowly raising the speed of the spacecraft to 225,000 mph before it begins to coast.[1] The 11,000-pound spacecraft, tentatively planned for launch around 2005, must be built to operate untended and continuously for decades (*Miami Herald* 1987b). TAU is a project for generations of astronomers.

TAU and *Voyager 2* are just two of many examples of efforts by NASA, industry, and university-based scientists to advance scientific knowledge through use and exploration of space. Much more needs to be done and will be done. Here are a few questions to which scientists are trying to find answers through space exploration and study:

- What laws of nature governed the birth and growth of the universe and now govern large-scale phenomena like the formation of galaxies, neutron stars, and black holes that cannot be duplicated in laboratories on Earth?

- How did the Sun, planets, satellites, and small bodies of the Solar System form? How have they evolved? Why are the giant planets so different from the terrestrial planets?
- How does energy flow from the interior of the Sun through its outer layers and into interplanetary space? How does it interact with the planets? How does the solar output vary? Does this cause Ice Ages and other changes in Earth's climate?
- What is the source of the Earth's magnetic field?
- What is the origin, evolution, and distribution of life in the universe? Are we alone?

There is enormous challenge to increase our scientific knowledge of the universe. As we increase our scientific knowledge, what appears impossible now may be commonplace in the next hundred years.

Exploring and Settling the Solar System. This is the fourth civilian use of space. We have already discussed exploring space to increase scientific knowledge of that frontier. In this section we address exploring space as a prelude to settling hospitable parts of the Solar System. At first glance, this suggestion seems preposterous until we recall that thousands of centuries ago, humankind did not inhabit North America or only dreamed of going to the moon.

Today, many scientists, such as Gerard O'Neill (1976), believe people will eventually colonize part of space. I believe these settlements will be small (30–100 people, mostly scientists and technicians), and primarily in Earth orbit, lunar orbit, and on the Moon. Many of the settlers will spend one to six months in space (doing mainly scientific studies), return to Earth, and then go back to space habitats. Many centuries into the future, these settlements may grow in population, agricultural production, robotic population for industrial production, and facilities for recreation, work, and waste management.

People may also continue exploring and prospecting space for minerals. Mars, the Moon, and many asteroids are believed to contain rich deposits of minerals, such as titanium, calcium, iron ore, and nickel, and others not found on Earth. Centuries from now, we may use those minerals for industrial, residential, and commercial use in space. The United States has plans to establish lunar bases, spaceports, and outposts on Mars or in Martian orbit during the twenty-first century. These plans do not include building habitats in space from minerals found in space, but they may eventually. Instead present plans call for building the parts of the structures (as lightweight as possible) on Earth and transporting them into space for assembly.

Exploring, prospecting, and settling the Solar System will be accompanied by many problems: spreading diseases in space, orbital debris and space pollution, wars over who owns particular parts of space, and conflicting political and economic ideologies in space.

Space Enterprise. Space business is already big business—over $25 billion annually in the United States alone. It consists of business physically located in space, such as communications and remote-sensing satellites, and supporting industries on Earth, such as the aerospace industry, space insurance industry, and payload processing business. Additionally space enterprise includes space transportation services, such as those provided by NASA's space shuttles, or by the European Space Agency's Ariane rockets.

In the future, space enterprise will grow larger. Here are some possibilities:

- Low Earth orbit (50–800 miles above Earth) touristic space travel, with stops at space stations and Earth observatories.
- Privately owned and operated space vehicles departing on frequent flights to orbiting space stations and (robotic) factories.
- Educational visits into space.
- New careers, such as space doctors and medical researchers. These professionals will research physiological effects of prolonged weightlessness on the human body and human adaptation to alien worlds and environments. Space doctors, space architects, environmental engineers, and human factors engineers will join together to design remote living and working quarters. Virtually every trade and discipline will be involved in space endeavors, from obstetrics to insurance.
- Products made in the microgravity environment of space, such as cancer-curing drugs, special lightweight alloys stronger than steel, and high-performance computer chips.
- Robotic factories in space making these and other products.
- A cadre of new professionals called space attorneys who deal with everything from product liability for space-made products to satellite collisions and malfunctions in space.

These and other possibilities will provide fascinating material for twenty-first-century (and beyond) business school textbooks.

Space business has markets on Earth, such as television broadcasting, weather forecasting, and military defense. Space business may also have markets in space, such as mining the Moon and asteroids (Koltz 1983), although the latter will not mature until the twenty-second century and beyond because of such obstacles as astronomical costs, limited economic demand, and technical barriers.

Space business will continue to generate tremendous employment opportunities in the aerospace industry, other firms with contracts from NASA space agencies in such states as Texas, Florida, Alabama, Colorado, and Virginia, and universities in these and other states that have space research grants and contracts (*Houston Chronicle* 1985b).

SUMMARY

There are two broad categories of uses of outer space: military and civilian. Some of the military uses of space are surveillance, attack warning and assessment, communication, and defense. Civilian uses of space include communication (e.g., radio, television, and telephone via satellites), advancing scientific knowledge through study of the universe, meteorology, space enterprise, and perhaps settling parts of the Solar System. Some of the military and civilian uses of space overlap, such as communication, meteorology, and defense. All of these uses of space are being pursued by the United States, as well as other nations, such as the USSR, Japan, and the PRC.

In the distant future, millions of years to come, humans may inhabit other planets. We may have space transportation vehicles capable of traveling at the speed of light (186,000 miles per second). Genetic engineering may be so advanced that people can live normally to 150 years old, and we may invent humanoid forms of life. If these advances occur, they will give rise to marketing opportunities for business worldwide, as well as ethical, cultural, and ideological controversies.

Space business—military and civilian—will continue to provide lucrative business opportunities for many American and foreign businesses for years to come in the areas of satellites, robotics, defense, scientific and industrial instrumentation, advanced computers and software for space R&D, remote sensing, and geodesy. Other kinds of business will be spin-offs of space technology, among them, space medicine, space law, and faster commercial transportation. American business must be poised to take advantage of these business opportunities and be aware of the obstacles and international competitors.

NOTE

1. TAU: thousand astronomical units. An AU is the 93-million-mile distance between the Earth and the sun.

CHAPTER 2

Economic Opportunities in and Obstacles to Space Commercialization: An Overview

The term *economic opportunities* is used in a broad sense. It encompasses economic opportunities for individuals (income, employment), companies (production, marketing, financing, insuring, and so on), countries (exports, imports, countertrade), and humanity (opportunities to enhance human welfare, quality of life, and standards of living). Some of these economic opportunities are abundant for business, such as those relative to the production and marketing of military, communications, remote sensing, and weather satellites. Others are futuristic, such as those that pertain to mining and colonizing the Moon (Koltz 1983), building lunar spaceports (*Pioneering The Space Frontier* 1986), and materials processing in space (MPS). This broad view of economic opportunities is implicit in the discussion that follows. The focus, however, is mainly on economic opportunities for firms and industry.

ECONOMIC OPPORTUNITIES IN SPACE COMMERCIALIZATION

On October 4, 1957, the USSR successfully launched *Sputnik I*, the first artificial satellite to orbit the Earth. The space age began. Since then, space business has grown tremendously and is now big business. Whereas the United States and the USSR were the major participants in the space race through the 1970s, today many other countries are seriously engaged in space commercialization for economic gain and defense purposes. These countries include Japan, the PRC, the thirteen-member nations of the

European Space Agency (ESA),[1] Canada, Brazil, India, and Italy. In the thirty years from 1957 to 1987, the nations of the world spent at least $300 billion on space (Dula 1985, p. 163) for such goods and services as satellites, space transportation vehicles, launch services, space-based defense systems, space stations, and space probes. Today these nations together spend over $100 billion annually on space business, with the United States and the USSR each spending about $25 billion annually on their space programs—equivalent to 50 percent of the world's space budget in 1988.

In terms of employment, thousands of people are directly employed in the space industry. NASA, which has an $11 billion fiscal 1989 budget, employs more than 24,000 people. Thousands more are employed by space agencies in other parts of the world, such as the USSR's Glavkosmos agency (the counterpart to NASA), Japan's National Space Development Agency (NASDA), the PRC's Ministry of Astronautics, and India's Department of Space. Many of these employees are top-notch scientists, engineers, computer hardware specialists and programmers, financial analysts, and medical doctors (space medicine). Many more thousands are directly employed globally by several aerospace and nonaerospace firms. These include Rockwell International (launch vehicles, MPS, space weaponry), McDonnell Douglas (MPS, spacecraft), 3M (Minnesota Mining and Manufacturing—computer crystals, films), General Dynamics (launch vehicles, space weaponry), Eastman Kodak (films), IBM (computers and software for space R&D), RCA (satellites), Aeritalia (the Italian aerospace contractor that built the Spacelab hardware for ESA), the German company Messerschmitt-Bolkow-Blohm (MBB), which makes space platforms called shuttle pallet satellites (SPAS), Marconi Ltd. (a British firm engaged in space defense research), British Aerospace (space transportation systems), Spot Image S.A. (France, remote sensing), and Mitsubishi Electric (Japan, satellites). (For a list of companies involved in space business, see Appendix B.)

Other firms are indirectly involved in space business. They include makers of space foods, providers of space insurance (e.g., Lloyd's of London, and Corroon & Black/Inspace, United States), and payload processors (e.g., Astrotech International). They also provide employment opportunities for many. An estimated 350 firms in the United States alone—aerospace as well as nonaerospace—have invested millions of dollars in space research, and some fifty others are negotiating joint research space ventures with NASA (NASA interview, Kuzela 1984). Some of these firms are 3M, John Deere, and McDonnell Douglas.

There are four points important to the discussion of economic opportunities in space commercialization. First, governments provide the groundwork for these economic opportunities and must continue to do so if space commercialization is to proceed. They must finance expensive technological developments that are beyond the normal means of corporations; oversee implementation; establish space laws; foster the overall climate for space exploration in which private enterprise must operate; and provide general leadership in space business. The central role of governments in space exploration has many historical antecedents; the Spanish government's support of Columbus in his voyages to the New World and the Portuguese government's support of Magellan in similar exploits are two instructive examples. As with these historical antecedents, the private sector will, with the passage of time, play an expanding role in space commercialization.

Second, private organizations that successfully engage in space commerce, such as IBM and RCA, are generally large, wealthy, international corporations with substantial track records, well-known expertise in their respective fields, and invaluable contacts with space agencies in the United States and other countries. Space business is largely the field of industrial giants that have staying power and a commitment. It is not a business for the faint of heart or for small, undercapitalized companies, although small companies can do well, as in payload processing and in the manufacture of special components for space transportation vehicles and space platforms.

Third, this chapter assumes a certain level of understanding to translate these broad business opportunities into more micro-opportunities for firms. For example, the production of communications satellites may be a broad business opportunity, but that opportunity spawns literally hundreds of smaller business opportunities that may fit neatly into the business manufacturing expertise and interests of many companies. These smaller business opportunities include making the solar panels, electronic components, and other parts of the satellite.

Fourth, economic opportunities do not fall into a company's lap. The company has to engage in serious strategic planning and business war games (James 1985) to earn its share of the economic spoils in space commerce. All this implies attention to producing quality products, and aggressive marketing to local space agencies and to contractors and subcontractors in space business. (See Appendix B for numerous examples of contractors, worldwide, in space business.)

For industry, the major economic opportunities in space commercialization can be classified into three categories: infrastructure needs, applications markets, and spin-offs of space technology. Infrastructure is necessary for the provision of other space products and services. It encompasses space transportation and launch services, in-orbit hardware and services, and ground support. The applications markets consist of satellite communications, MPS, remote sensing, and defense—that is, some aspect of space is critical to the provision of service. Satellite communications, defense, and remote sensing profit from the vantage point afforded by space; MPS utilizes other physical attributes of space (most notably microgravity) to produce special alloys, pharmaceuticals, and other materials that cannot be made on Earth. Spin-offs are self-explanatory.

Infrastructure Markets

Transportation and Launch Services. Transportation is imperative for the commercial development of space. It is also the most mature of space ventures. It consists of unmanned space probes, such as *Voyagers 1* and *2* launched in 1977, and the USSR's *Vega* launched in December 1984 to Venus and Halley's comet; manned explorations via space shuttles, such as the *Atlantis*, *Discovery*, and *Columbia* ($2 billion each); minishuttles, such as ESA's *Hermes* ($4 billion); space planes being developed by the USSR, and West Germany (the *Sanger*); and the launching of satellites by countries such as the United States, USSR, and Japan and for firms such as AT&T and RCA. Japan, the PRC, and Brazil also have space transportation vehicles, and the British are developing an advanced rocket plane, *Hotol* ("horizontal takeoff and landing").

A U.S. space shuttle costs from $3 billion to $4 billion to build (in 1988 dollars) and requires several million dollars more per year for maintenance and operation. IBM, RCA, General Electric, and Rockwell International are just a few of the many corporations that build parts for space shuttles. Rockwell International, the major builder of the U.S. space shuttles, also has a multimillion dollar contract per year to service the shuttles (Byars 1985a).

There is tremendous economic competition among the United States, ESA, and the PRC for commercial launch business. The USSR entered that business in 1987. Both the United States and ESA charge about $25 million each to launch a single satellite. The United States does so on its manned space shuttles and ESA on the unmanned *Ariane* launch vehicles. The PRC and the USSR charge about $20 million for similar services. Each *Ariane 4* currently costs about $80 million, and Arianespace (the commercial arm of ESA) recently ordered fifty of them, which will keep it stocked up until 1998 (*Economist* 1987a). The heavy-lift *Ariane 5* now being developed will be able to lift about 6 tons up to geostationary orbit (22,300 miles above the equator) or about three satellites (as against 4 tons for the *Ariane 4*) and will cost approximately $3.1 billion each.

In the aftermath of the *Challenger* tragedy of January 28, 1986, many firms and countries are finding renewed economic opportunities in building and marketing expendable launch vehicles (ELVs). These companies and countries include General Dynamics (*Atlas/Centaur* rockets), Martin Marietta (*Titan* rockets), McDonnell Douglas (*Delta* rockets), ESA (*Ariane* rockets), Japan (H-Series rockets), and the USSR (*Proton* rockets, and the newer, more powerful *Energia* rockets).

In-Orbit Hardware and Services. Space stations and space platforms are typical in-orbit hardware. They provide in-orbit services like workspace and utilities for MPS, satellite repair facilities, scientific research facilities, spacecraft refueling, and instrumentation for a variety of space science, astronomy, physics, and Earth observation experiments. The USSR's *Salyut 1* through 7, and *Mir* ("*Peace*") are examples. The U.S./International Space Station now being developed by NASA, ESA, Canada, and Japan at a cost estimated at between $20 billion and $25 billion is another example.[2] This space station will be about the size of a football field, will weigh around 40 tons, and will be assembled in space via twenty to twenty-five shuttle flights during 1995 to the year 2000. It will have several interconnected modules for power, work areas, and living quarters for about six to eight persons. It will be in low Earth orbit (LEO), around 300 miles above Earth.

The U.S./International Space Station has given birth to numereous business opportunities worth millions of dollars for architectural and engineering design work and contracts to build the living quarters, power systems and laboratories, and waste management facilities on the station (Payne 1987). Among the major contractors vying for large billion-dollar contracts to design and build portions for the space station are Boeing, Rockwell International, Grumman Aerospace, Ford Aerospace, Teledyne Brown Engineering, Martin Marietta, General Electric, TRW, Hughes Aircraft, United Technologies, Rocketdyne, McDonnell Douglas and General Dynamics (Payne 1987, p. 126).

In addition, there are several examples of unmanned and man-tended space platforms now being developed by various firms and countries. They include ESA's European Retrievable Carrier (EURECA), more advanced derivatives of the EURECA called the EUROS (European Automated Platform), the SPAS, and the Industrial Space Facility (ISF) being built by the Houston, Texas, firm Space Industries, Inc. Some space platforms, such as ISF, are being designed to be integrated with the U.S./International Space Station. Generally these platforms will vary in size from about one-sixteenth to one-quarter the size of a football field and be able to carry various payloads. They will cost $300 million and up, depending on size and onboard technological instruments and facilities.

Governments will use their space stations for scientific work, MPS, and military surveillance. Private sector makers of space platforms, however, face a monumental task of marketing their space platforms to the private sector. Problems include high lease costs of workspace, instruments, and utilities (estimated at between $2 million to $4 million per month per space platform), high round-trip space transportation costs of $10,000 per kilogram for LEO transportation (*Economist* 1984b), unreliable transportation schedules, the space naiveté of many potential users of space platforms, the expense of space ventures, difficulties in obtaining financing from the capital markets, difficulties in obtaining space insurance, especially in the aftermath of the *Challenger* tragedy, and the absence of developed markets for MPS products.

Ground Support. Ground-based support includes the following: (1) preparation and processing of payloads for flight (e.g., checking the electronic components of a satellite before loading for launch), as done by Astrotech International (Cape Canaveral, Florida); (2) the manufacture of components for the U.S. space shuttles (by Rockwell International and General Electric, for example); (3) the manufacture of space suits, space foods, and other products; (4) the manufacture of special on-board computers and others for mission control stations (by IBM, for example); and (5) the provision of space insurance.

In the United States alone, the ground support industry takes in $1 billion to $2 billion per year; payload preparation and processing costs are $750,000 to $1 million for a single communications satellite; and the space suit worn by a U.S. astronaut for work outside the spacecraft costs around $2 million each. (Suits worn inside the spacecraft are much less expensive or cumbersome. They are called flight suits.) ILC Dover, Inc., Delaware, is a manufacturer of space suits (McGinley 1987b).

Applications Markets

Satellites. The manufacture and marketing of various types of satellites (e.g., communications, remote sensing) and the invaluable services they provide, such as transmission of television and radio broadcasts, telephone conversations, and electronic mail and other data, represent some of the best (and among the first) economic opportunities in the space applications market. In fact, the satellite industry is the most mature of the industries currently involved in space commercialization. Each satellite typically costs from $50 million to over $250 million, depending on size, number of transponders, and use of advanced satellite technology. (Most communications satellites being built today have an average of around twenty to twenty-five transponders, but some have as many as fifty-five. Businesses often lease a transponder for about $1 million to $2 million per year.)

Today communications satellites are the most lucrative commercial payload (cargo) being transported into space. Some are for countries such as the United States, Japan, West Germany, Australia, Mexico, the PRC, Indonesia and Brazil. Others are for commercial firms like AT&T/COMSAT (Comstar series), Hughes Communications (Galaxy series), RCA (Satcom series), GTE (Spacenet series), and Ford Aerospace (Fordsat series). RCA and Hughes Communications are two leading U.S. satellite manufacturers. Together they account for more than 60 percent of U.S. satellite production. The Soviets, Japanese, and Chinese also build sophisticated satellites.

Since the Soviet Union launched *Sputnik* in 1957, about 3,000 satellites have been orbited by various countries and firms; 90 percent of them belong to the United States and USSR. In recent years, the United States has been averaging somewhere around ten to fifteen satellite launches per year and the USSR about a hundred (generally smaller satellites). Two-thirds of both countries' satellites are for military communications purposes, and the other third primarily transmit radio broadcasts, telephone conversations, electronic mail, weather information, and other data.

Annual gross revenues for the communications satellites segment could be as high as $40 billion to $100 billion in the United States alone by the year 2000 (Osborne 1985, p. 51). Annual gross revenues from the sale of remotely sensed data could reach $2 billion by the year 2000. Remote sensing could be a major growth industry in the twenty-first century and beyond, with major customers for remotely sensed data being governments and the minerals, forest, and maritime industries. In the United States, government agencies that use remotely sensed data include the U.S. Department of Agriculture and the U.S. Department of Defense. The satellite industry also creates other product and service opportunities for firms, such as the need for earth stations, antennas, computers, and electronic instrumentation.

Manufacturing and Materials Processing in Space (MMPS). The environment of space—characterized by an absence of vibration, near-perfect vacuum, sterile environment, unfiltered sunlight, and, especially, microgravity—provides a potentially valuable laboratory for experimentation and processing of certain chemicals, pharmaceuticals, and alloys that may be produced more efficiently and in higher quality in space than on Earth. John Deere is studying zero-g (zero-gravity) iron processing to improve its cast iron foundries; McDonnell Douglas Corporation is experimenting with production of interferon (a drug that blocks or inhibits the growth of cancer cells) and other medicines in space; and 3M is studying organic and polymer chemistry in zero-g to improve its plastics and adhesive products. Materials processing in space (MPS) has been conducted on the U.S. *Skylab* (the first U.S. space station), in the U.S. space shuttles, on *Salyut 1* through 7, and *Mir*. MPS will also be conducted on the U.S./International Space Station and on space platforms.

The pharmaceutical industry is likely to be the first user of space for manufacturing, using the weightless environment for the delicate separation of complex, nearly identical substances. Jastrow (1984) states that some rare and expensive medicines could be made more purely, cheaply, and efficiently in space than on Earth using electrophoresis.[3] These medicines include urokinase, an enzyme that dissolves blood clots, used to treat victims of pulmonary embolism and heart attacks caused by blood clots and costing about $1,000 a dose; Factor VIII used in treating hemophiliacs and currently priced at $3,000 a dose; and the beta cell, normally produced in the pancreas and required for the production of insulin for treatment of diabetes. Manufacturing would be done by robotic machines and by scientists from Earth working on space stations and space platforms for a few months at a time. Shuttles or space planes would pick up the scientists, as well as processed products, and return them to Earth. This has already been accomplished on a small scale by the Soviets (*Miami Herald* 1987a).

Defense. The defense component of space commercialization could be classified as either part of the space infrastructure market or the applications market. Defense weaponry (e.g., lasers) and surveillance systems in space represent the most potentially lucrative of all the economic opportunities in space commercialization. Several nations spend billions of dollars each year on a variety of defense systems in space, consisting primarily of military and communications satellites and accompanying space weapons. In 1985, space weaponry received an additional shot in the arm with President Reagan's proposed Strategic Defense Initiative (SDI), essentially a space-based antimissile defense system with ground-based support. The United States has awarded SDI research contracts to Britain, one for $8.7 million (*Miami Herald* 1986); West Germany; and Israel. The USSR is also conducting its own SDI and counter-SDI research. SDI research in the United States by defense contractors, NASA, and universities is variously estimated at between $5 billion and $10 billion a year. One estimate is that SDI will cost up to $2 trillion for full deployment around the year 2000 (*Higher Education Advocate* 1987).

The defense component of space commercialization is here to stay. The United States and USSR and their allies represent a ready market for space weaponry in spite of laws in both countries restricting the export of sensitive, advanced, and secret defense systems. The major beneficiaries of the defense component of space commercialization are the defense contractors. In the United States, billions of dollars worth of defense contracts are awarded each year to General Dynamics (the largest defense contractor in the U.S.), Martin Marietta, Hughes Aircraft, Rockwell International, General Electric, and TRW. Similarly, the British Ministry of Defence, ESA, and India's Department of Space award large defense contracts to local corporate contractors. British Aerospace and Marconi Ltd., are two examples of defense contractors in the United Kingdom.

Spin-offs of Space Technology

Spin-offs inadvertently tie space business to Earth business and have created market opportunities in space commercialization (Goodrich, Kitmacher, and Atmey 1987, pp. 81–82).

The X-ray inspection systems that examine luggage at airports were first developed for astronomical satellites studying celestial X-ray sources. American Science and Engineering, Cambridge, Massachusetts, the company that developed the highly sensitive X-ray detectors for NASA, now makes the airport detectors.

NASA asked Black and Decker to make a rechargeable, cordless drill that astronauts could use to drill moon core samples. The company thought it was such a good idea that it utilized this new technology to develop an entire line of cordless power tools, now popular among home builders, construction workers, and home handymen.

U.S. sports fans at the Silver Dome in Pontiac, Michigan, the Astrodome in Houston, Texas, and other facilities with similar fabric domes owe their comfort to technology used to make astronaut space suits. These domes are made of Teflon-coated fiberglass fabric, which was developed by Owens-Corning Fiberglass Corporation to protect astronauts from the hostile environment of space.

Advanced computer hardware and computer programs that IBM developed for NASA to monitor and control spacecraft are now being used in automated industrial manufacturing.

Car fenders, bumpers, and many other products are being made of a new class of composite materials that are stronger but lighter than metal. These composites, made of fiber-reinforced plastics, were originally developed for spacecraft construction.

New meteorological forecasting techniques, developed jointly by NASA and the U.S. Air Force, are used in predicting weather conditions.

In addition to these spin-offs, whole new industries—including computer science, solid-state electronics, and communications satellites—were advanced as a result of NASA's research and development in collaboration with some U.S. industrial giants. New jobs and economic growth have resulted from space business. Space business has created new frontiers in business on Earth.

In order for businesses to keep abreast of space spin-offs and take advantage of market opportunities they present, they must keep in contact with their respective space agencies and patent offices and read journals that report some of these spin-offs, such as the *Economist*, the *Wall Street Journal*, *Technology Review*, *Aviation Week & Space Technology*, and *Tech Briefs* (a monthly magazine published for NASA by Associated Business Publications, New York, New York, that covers many aspects of the U.S. space program). (See Appendix A for other potential sources of information on spin-offs.) Generally *Aviation Week & Space Technology*, published weekly by McGraw-Hill, is the best source of reliable information on the space industry. NASA also publishes annually a booklet of its major space technology spin-offs. A copy of the 1986 issue of this booklet, titled *Spinoff 1986. NASA*, is available, as other annual editions are, from the Superintendent of Documents, U.S. Government Printing Office, Washington, D.C. 20402.

The major drawback of the published sources of information on spin-offs, at least from a business perspective, is that these sources usually discuss spin-offs already under license from NASA, so they are often lost business opportunities for other businesses. NASA, however, grants both exclusive and nonexclusive licenses for production and marketing of its spin-offs.

It is important to understand that although many spin-offs are the property of NASA, many others are not. That is, many major contractors working with NASA have clauses in their contracts stipulating in effect that spin-offs they develop in their work for or with NASA are their sole property. Such a contractor is free to develop and market the spin-off. Many corporations, including 3M, IBM, Lockheed, and McDonnell Douglas, are thereby doing their own R&D on their own spin-offs. Corporate R&D on spin-offs is usually kept as proprietary or confidential information for obvious reasons.

Other Miscellaneous Economic Opportunities

Other miscellaneous economic opportunities in space commercialization include the contracts awarded by space agencies and corporations to organizations to do advance or feasibility studies. NASA has already spent $600 million for advance studies on the U.S./International Space Station, and Boeing Company, which heads a group of five companies vying for a $2.5 billion contract to build the crew quarters and an adjoining laboratory, has spent $70 million preparing its proposal (Payne 1987). Similar sums have already been spent by some of the largest contractors in the world, among them General Dynamics and McDonnell Douglas, for proposals for the space station, space platforms, and other space infrastructure needs.

THE BEST ECONOMIC OPPORTUNITIES IN SPACE

The best current economic opportunities in space commercialization are defense systems, space transportation systems (STSs), artificial intelligence, robotic technology, and remote sensing.

The defense business is large-scale, lucrative, and here to stay in the United States, USSR, Britain, West Germany, Japan, Israel, and their allies. In the United States, large defense contractors such as General Dynamics and Hughes Aircraft enjoy billions of dollars worth of defense contracts each year from the U.S. government. In Britain, British Aerospace and Marconi Ltd. are examples of successful defense contractors.

STSs are imperative if space commercialization is to proceed and succeed. They are a necessary part of the space infrastructure. STSs include space shuttles, space planes, space probes, ELVs, orbital maneuvering vehicles (OMVs), "tugboats," and a variety of spacecraft designed for different purposes. In the aftermath of the *Challenger* accident, there is renewed emphasis on the manufacture and marketing of ELVs by producers like McDonnell Douglas (*Delta* ELVs) and General Dynamics (*Atlas/Centaur* ElVs).

Artificial intelligence, mainly in the form of computers, is essential for space commercialization. Computers are core components of all spacecraft and of ground-support systems. They help to direct, control, and monitor spacecraft and their internal systems and are necessary for data management. The industrial giant in the area is IBM. Artificial intelligence is a growth field of tremendous importance in space industry and space science.

Robotics will also be a tremendous growth area. There will be a need for robots to explore outer space prior to on-site exploration by humans. Similarly, future MPS will require robotic factories. (There are robotic factories in the United States and Japan so advanced as to operate one or two shifts out of three without human attendance.) Space robotic factories will have several basic components for industrial production: transporting machines, processing machines, and artificial intelligence. Japan is one of the leading countries in the production and marketing of industrial robots used for spot welding, automobile painting, and industrial manufacturing. About 40,000 industrial robots are in use in Japan, compared to some 10,000 in the United States.

Remote sensing will also grow in importance. As these economic opportunities flourish, so will ground support. Most of these economic opportunities, however, require large amounts of capital, commitment to space commercialization, and staying power. Therefore large, wealthy corporations are generally more able than small companies to exploit commercial opportunities in space.

OBSTACLES TO SPACE COMMERCIALIZATION

The path of space commerce will be strewn with many obstacles—some internal to the firm, some external, some both.

Costs

Space business is expensive business. It is estimated that the average start-up cost for a manufacturing plant engaged in nonspace business is around $5 million; a similar operation for a space venture is $50 million and up. Typical costs include huge sums of borrowed capital, large interest expense, high salaries for high-tech scientists and engineers, plant and equipment costs on Earth, transportation into space, the cost of space factories, and so on (Shifrin 1984). So long as these costs remain high—and they generally will—space business will grow very slowly. Consequently, space business will continue to be the turf of wealthy industrial giants for many decades. (Many of them are listed in Appendix B.) Small businesses will generally be outflanked by large firms.

Profit Squeeze and Long Payback Periods

The corollary of astronomical costs of space ventures is that profits will generally be slim or nonexistent for many years (sometimes ten or more).

Such long payback periods are anathema to American business or alien to the American business culture. Executives in space ventures have to learn to think in terms of long payback periods. During such times, technological advancements, changes in government policies, and other uncertainties of high-risk space business are likely to render some space investments fruitless.

Space Transportation Problems

Transportation into space is the most essential component for the commercial development of space. Without it, the space frontier cannot be developed. Some development can be accomplished through the use of space probes, ELVs, and the like, but full commercial exploitation of space requires human beings. Human beings in space are superior to robots for observation, creativity, and the application of judgment and will always be, in spite of the development of artificial intelligence and expert systems.

But transportation problems plague the U.S. space industry, especially in the aftermath of the *Challenger* tragedy. These transportation problems include high cost, integration difficulties with NASA's space transportation schedules, frequent transportation delays as a result of mechanical problems with the space shuttles, and weather problems. Space shuttle customers cannot operate their space business under such uncertainties. All this may give rise to private sector development of space transportation services in the twenty-first century. In fact, there is discussion at NASA headquarters in Washington, D.C., about privatizing space transportation. (For a general overview of privatization in America, see Goodrich 1988.)

Markets and Marketing

The absence of already developed markets, the uncertainty about future markets, and the gargantuan task of space marketing together represent one of the biggest obstacles to space business. For example, will pharmaceuticals capable of being effectively produced in space find a large enough market on Earth? Or can Earthbound biotechnology techniques do just as well? Markets are likely to exist for products with a high value-to-weight ratio, such as electronic devices, specialty glasses, alloys, some pharmaceuticals (Deudney 1982), and space weaponry.

Technological and Engineering Barriers

Technological barriers, another major obstacle to the commercial development of space, generally are caused more by unfamiliarity with hardware design for space use than by the difficulty in designing the actual hardware. The technological barriers will retard development of the space industry, including MPS, other orbital operations, and space transportation.

Space Insurance Problems

In the 1980s, many firms have had enormous difficulties obtaining space insurance, especially in the aftermath of the failure of several shuttle and *Ariane*-launched satellites, the loss of U.S. *Delta* and *Atlas/Centaur* rockets, the failures of a few *Ariane* rockets, and the destruction of the *Challenger*. In fact, the space insurance industry has temporarily stopped issuing new policies. The industry has lost more than $600 million worldwide during the 1980s. Firms in the space business are reluctant to build space hardware when they cannot obtain insurance for their expensive assets in space or at reasonable costs. (See chapter 9 for more details on space insurance.) Space insurance difficulties have therefore retarded the commercial development of space.

Other Obstacles

Among the many other obstacles to space commercialization are competition, complex government regulations, time delays, inadequate patent protection, safety concerns, inertia, and the complex political, social, military, monetary, and legal challenges that must be dealt with on a national and international scale. These obstacles are not meant to suggest that we abandon space commercialization; they are meant to alert industry to the realities of space business. They also pose challenges to industry and are the catalysts for technological progress.

MARKETING IMPLICATIONS

The marketing implications of the opportunities in and obstacles to space commercialization are both micro and macro in nature. They pertain to joint ventures, antitrust law, space law, multinational cooperation and marketing, new products and new markets, advertising, and the emergence of a more marketing-oriented NASA.

Joint Ventures

Large companies in the space industry will enter into joint ventures with other firms or with their governments in order to defray the heavy capital outlay that space ventures entail. These joint endeavors will spawn new collaborative marketing efforts hitherto unheard of among large firms. 3M, General Motors, Kodak, and other nonaerospace industrial companies are discussing how they can move as a group to exploit commercial space opportunities (*Aviation Week & Space Technology* 1985b). NASA also has some fifty joint endeavor agreements (JEAs) with U.S. companies, including Honeywell, (defense industry materials), Du Pont (catalytic materials),

McDonnell Douglas (pharmaceutical production in space), Arthur D. Little (research on long-term blood storage), and Space Industries (construction of an industrial space facility for lease). NASA provides payload preparation and transportation into space at little or no cost to the client in return for the knowledge and research gained by the client in space experimentation.

Antitrust Laws

The emergence of joint ventures and collaborative marketing will require modifications of antitrust laws in the United States and perhaps in other parts of the world. These laws affect the marketing of goods and services internationally and, perhaps in the near future, extraterrestrially.

Space Law

Space law, in its infancy, will evolve as have the laws of the sea: slowly, tediously, and by trial and error. Space laws—covering patent rights, space rights, and labor-management relations—will fashion the marketing of products and services for space use. (See chapter 10 for more details on space law and space commercialization.)

Multinational Cooperation and Marketing

In order to meet the complexities of space commercialization—legal, economic, political, and social—multinational collaboration and marketing will grow in importance. The collaboration of the United States, the United Kingdom, Israel, and other countries in SDI research is a case in point.

New Products and New Markets

Possibilities for new products and services and new markets spawned by space exploitation include touristic travel to LEO for around $50,000 per person for a twelve-hour trip (*Aviation Week & Space Technology* 1985o), faster and safer commercial aircraft, new space-made pharmaceuticals as cancer cures, advanced expert systems and artificial intelligence for industrial and scientific uses, advanced radar, supercomputers, and further development in fiber optics communications systems. Rockwell International, for example, is doing work on expert systems and GTE on fiber optics.

Companies must be poised to take advantage of new product possibilities and new markets by keeping abreast of NASA, NASDA, and other space agencies, the space commercialization literature (see Appendix A), relevant conferences, and government patents that are spin-offs of space R&D.

Advertising

To attract customers of space products, space businesses will increase their advertisements in technical journals, such as *Space Commerce*, *Aviation Week & Space Technology* and *Space Business News*. They will also advertise industrial and consumer goods in more traditional media.

A More Marketing-Oriented NASA

A welcome implication of the numerous opportunities in and obstacles to space commercialization is that NASA is becoming more marketing oriented rather than primarily engineering or production oriented. In 1985 the Office of Space Commercialization was established at NASA Headquarters in Washington, D.C.; more JEAs between NASA and private corporations are being made; and more seed money is going from NASA to universities and private industry to help them fund space R&D. Other government departments, such as Transportation and Commerce, are becoming more involved in space legislation and law enforcement as well.

SUMMARY

Space business is big business. Globally it involves over $100 billion annually—for satellites, space transportation, and launch vehicles, space transportation services, defense, and so forth. Several countries are involved: the United States, the USSR, the member countries of ESA, Japan, the PRC, Italy, Brazil, and India. Their space agencies play a central role in advancing the interests of space commercialization. These agencies include NASA, NASDA, ESA, Glavkosmos, the Ministry of Astronautics in China, and India's Department of Space.

Current economic opportunities in space commercialization can be classified into three main categories: infrastructure market needs, applications markets, and spin-offs of space technology. Examples of infrastructure needs are space transportation vehicles, launch services, and space stations. The applications market includes satellite communications, defense, and MPS. Spin-offs include X-ray inspection systems at airports, cordless power tools, fabric domes at sport arenas, and miniaturization in general. Some of the best economic opportunities in space are defense, ELVs, robotics, remote sensing, and artificial intelligence.

Generally space commerce faces many obstacles: tremendous transportation costs, extraordinary capital needs, technological barriers, insufficient economic demand in some cases, uncertain markets, difficulties in obtaining space insurance, and safety problems. As time goes by, many of these problems will be solved as governments and world corporations become increasingly involved in space commercialization.

The marketing implications of these opportunities in and obstacles to space commercialization include more emphasis on medical research relative to the space environment; the growth of joint space ventures among governments and corporations to defray the heavy capital outlays that space ventures entail (between NASA and U.S. corporations, for example); more collaborative, multinational space ventures (the United States and UK in SDI research, for example); the emergence of new international space laws (product liability, tort and criminal law in space); and growth in symbiotic marketing (Varadarajan and Rajaratnam 1986; Adler 1966). Additionally, many economic opportunities in space commerce will not really take off until after the U.S./International Space Station is fully operational (in the first decade of the twenty-first century) and after other space infrastructure needs, such as more space transportation vehicles and other space stations, are in place.

Nevertheless, by 2088 A.D., space commerce will achieve bold accomplishments. These will include touristic travels to LEO, lunar spaceports, space stations in different orbits, manned journeys to Mars, advanced space-dependent communications and information systems, and technologically superior space probes that will photograph, map, and document the resources of near and distant space. Through advances in space medicine and genetic engineering, humans may some day live normally to one hundred and fifty years. We may also discover that we are not alone in the universe.

Space commercialization is here to stay. The true believers today may become the industry giants of tomorrow. Space commercialization and industrialization is another step in human evolution.

NOTES

1. ESA, formed in 1975, consists of Belgium, Denmark, West Germany, France, Ireland, Italy, the Netherlands, Spain, Sweden, Switzerland, Britain, Norway, and Austria.

2. Note, however, that the $20 billion to $25 billion cost is construction costs only; it does not include launch, operation, and maintenance costs. Twenty-five round trips of the U.S. space shuttle, carrying space station construction materials and crew, would cost about $6 billion. NASA maintains that it can support the space station with eight shuttle flights per year. One estimate is that this will cost an additional $2 billion per year. A lifeboat will also have to be developed to return the crew to Earth in case of an emergency. This vehicle will cost around $2 billion. Therefore, the initial space station budget requirement is about $30 billion to $35 billion.

3. Electrophoresis is important in biological research. In electrophoresis, proteins, enzymes, and other molecules are separated by means of electrical charges. Such separation is fundamental to biological research. Electrophoresis does not work

reliably in Earth's gravity field. Samples of dense material, for example, tend to collapse in a blob. McDonnell Douglas reports that one experiment in space produced as much as 716 times more separated product than its equivalent on Earth would have, with a fourfold to fivefold improvement in purity.

CHAPTER 3

Space Operators

The principal operators in outer space are private entrepreneurs, government agencies, and international agencies.

ROLE OF PRIVATE ENTERPRISE

Private enterprise, which plays a significant role in space commercialization, takes three forms: incorporated companies, partnerships, and sole proprietorships. These enterprises manufacture and market space products and provide many services, such as space insurance, satellite broadcasts, and payload processing. They are essentially independent of government in their policy and management, and their objectives are primarily commercial. However, the country(ies) in which these entrepreneurs operate establishes space laws, regulations, and procedures that circumscribe their activities (just as other governmental laws establish operating guidelines for nonspace entrepreneurs).

Many companies, such as IBM, 3M, and General Electric, have branches or subsidiaries in their own and other countries. The parent company and its subsidiaries are subject to the law and jurisdiction of the country in which the parent company has been incorporated and of the country in which the subsidiary is located. Companies also form consortia (the consortium being for some purposes the effective operator), or they engage in joint ventures, causing further problems concerning jurisdiction to arise. Examples of joint (space) ventures are McDonnell Douglas and Ortho Pharmaceutical (pharmaceutical production) and 3M and NASA (materials processing in space).

Table 3.1 lists some of the major companies in the United States—aerospace as well as nonaerospace—that are engaged to varying degrees in space commercialization. They include IBM, General Electric, General Motors, Martin Marietta, 3M, and Satellite Business Systems (SBS). SBS, a tripartite enterprise of IBM, the Communications Satellite Corporation (COMSAT), and Aetna Life Insurance Company, is engaged in the manufacture of satellites and in setting up a network of receiving information service through satellites. A number of other companies (e.g., AT&T) have been authorized by the Federal Communications Commission (FCC) to provide mixed satellite services and are regarded as common carriers.

Many companies and countries outside the United States engage in space business. The leading countries are the USSR, countries participating in ESA, Japan, the PRC, Israel, and Canada. Less active participants are Mexico, Brazil, Indonesia, Australia, India, and Pakistan. A few examples of non-U.S. companies engaged in space commercialization are British Aerospace in the United Kingdom, OTRAG (Orbit Transport and Rakaten) in the Federal Republic of Germany (FRG), Spot Image (France, remote-sensing services), and Great Wall Industrial Corporation (satellite launches) in the PRC. (See Appendix B.)

There is a great deal of collaboration among U.S. and foreign companies through consortia and joint ventures. British Telecom International, for example, is in partnership with SBS on part of the business information service; and companies in the UK, FRG, Israel, and the United States are collaborating on SDI projects.

There are also examples of satellite communication corporations established by government legislation that range from public agencies to semicommercial enterprises. Two examples outside the United States are the USSR Maritime Satellite Communication Association and Telesat Canada. The former is like a public agency and has all-Union functions and legal personality. (The title *Association* is equivalent to *society* or *company*.) Telesat Canada, which began operation in 1969, is a corporation established by the Canadian Federal Parliament as a national satellite telecommunications system.

Inside the United States, the most notable example of a satellite communications corporation established by government legislation is COMSAT. COMSAT was set up when the U.S. Congress enacted the Communication Satellites Corporation Act of 1962. The broad objective and international mandate of the Act was to establish a commercial communications satellite system in conjunction with other countries and as part of an improved global communication network to contribute to world peace and understanding. COMSAT was authorized by the Act to:

1. Plan, initiate, construct, own, manage, and operate itself or in conjunction with foreign governments or business entities a commercial communications system.
2. Furnish, for hire, channels of communication to U.S. communications carriers and to other authorized entities, foreign and domestic.
3. Own and operate satellite terminal stations when licensed by the FCC.

COMSAT was a U.S. corporation, created by statute and governed by U.S. law. It was also required, though in rather imprecise terms, to submit to presidential supervision and to seek advice from the Department of State because of the international range of its functions.

In 1965, COMSAT launched *Early Bird* (also called *INTELSAT I*), the first commercial communications satellite. Since then, it has launched many advanced satellites and created a worldwide communications system encompassing the transmission of telegraph, telephone, television, and other forms of communication across oceans and continents.

Attitudes and policy toward the role of private enterprise in space vary. In the United States, there is a movement toward expanded private sector investment and involvement in space activities. In fact, this is just a small part of what appears to be a growing privatization movement in the United States (Goodrich 1988; Roth 1985; Hanke 1984; Grace 1984). In Japan, there is close cooperation between the government and private enterprise in space ventures. In the USSR, the government plans, initiates, and runs space activities.

There is a high level of collaboration between the private sector and public agencies in the United States, evidenced by the growing number of JEAs and memoranda of understanding (MOU) between NASA and numerous private companies such as IBM, 3M, McDonnell Douglas, and General Dynamics (Kozicharow 1984b, p. 98). The private sector is becoming more and more involved in space activities than it was years ago; its impetus for this increased involvement is profits.

GOVERNMENT AGENCIES

Perhaps the best-known government agency dealing with space is NASA, which works closely with the U.S. Department of Defense and the U.S. Department of Commerce, as well as with U.S. corporations interested in space ventures. Important space agencies outside the United States are the USSR's Glavkosmos, Japan's NASDA, the British National Space Center, the PRC's Ministry of Astronautics Industry, and India's Department of Space.

Generally these agencies help to formulate and execute space policies and activities in their respective countries. Except in the general case of the USSR and PRC, these agencies also collaborate a great deal with the private sector

Table 3.1
Some U.S. Firms Involved in Space Commercialization

Company	Space Activities
Aeros Data Corp., Bethesda, Maryland	Software to analyze remote-sensing data
Aluminum Co. of America (ALCOA) Pittsburgh, Pennsylvania	Aluminum
American Science and Technology (AS&T) Bethesda, Maryland	Remote-Sensing Systems
Astrospace, Santa Clára, California	Computer Software, Getaway Specials support
Astrotech International Pittsburgh, Pennsylvania	Processing shuttle payloads
Ball Aerospace Boulder, Colorado	Remote-sensing satellites and services
Batelle Columbus Laboratories Columbus, Ohio	Remote-sensing, and data management
Bethlehem Steel Bethlehem, Pennsylvania	Metal alloys
Center for Space Policy Cambridge, Massachusetts	Consulting
Device Engineering Alexandria, Virginia	Provides Getaway Specials support services, e.g., batteries, recorders, cameras
Earth Observing Satellite Co. (EOSAT) Arlington, Virginia	Remote-sensing
Eastman Kodak Rochester, New York	Films, glass alloys
Ecosystems International Millersville, Maryland	Consulting, space processing experiments
General Electric Corporation Fairfield, Connecticut	Electrical components for U.S. space shuttles, and the U.S./ International Space Station

Table 3.1 (Continued)

Company	Space Activities
General Dynamics San Diego, California	Satellite launch vehicles, space weaponry
General Motors Research Laboratories Warren, Michigan	Knowledge-building microgravity research experiments in space, e.g. combustion physics and lubricants
Getaway Special Services Bellevue, Washington	Getaway payload design and integration services
Grumman Corporation Houston, Texas	Data systems,alloys
Hewlett Packard Palo Alto, California	Computer software, genetic engineering
Hughes Aircraft Houston, Texas	Satellites
Honeywell, Inc. Lexington, Massachusetts	Computer parts, remote sensing
IBM Houston, Texas	Computers, software for space R&D
John Deere & Co. Moline, Illinois	Metal alloys, metallurgy
Lockheed Space Operation Cape Canaveral, Florida	U.S. space shuttles
Lovelace Medical Foundation Albuquerque, New Mexico	Antibody research for treatment of cancer and other crippling diseases
Martin Marietta Bethesda, Maryland	Rockets
Microgravity Research Associates Coral Gables, Florida	Crystal growth in space

Table 3.1 (Continued)

Company	Space Activities
Morton Thiokol, Inc. Brigham City, Utah	Rocket boosters, defense systems
3M Corp. (Minnesota Mining and Manufacturing) St. Paul, Minnesota	Organic crystals, thin films, plastics
Orbital Sciences Corp. Vienna, Virginia	Manufacture of space shuttle components and satellite launch vehicles
Radio Corporation of America (RCA) New York, New York	Satellites
Satellite System Engineering Bethesda, Maryland	Consulting
Space Industries, Inc. Houston, Texas	Developing the Industrial Space Facility, a free-flying scientific and industrial space platform (mini-factory)
Space Vector Northridge, California	Space transportation systems
Sparx Corp. New York, N.Y.	Commercial remote-sensing services
Spot Image Corp. (Subsidiary of France's Spot Image S.A.) Washington, D.C.	Remote-sensing services
Union Carbide/Oak Ridge National Laboratories Oak Ridge, Tennessee	Glass-forming alloy systems
Westinghouse Corp. Baltimore, Maryland	Ultrapure, bubble-free glass

in space R&D. It is of paramount importance for the private sector to keep in touch with the space agencies in their respective countries for contracts, policy information, and so on.

REGIONAL AGENCIES

Four of the most important regional agencies involved in space activities are INTELSAT (International Telecommunications Satellite Organization), Intersputnik, ESA, and ARABSAT (Arab Satellite Telecommunications Consortium).

INTELSAT is an international agency that provides satellite communications between nations. It has fifteen satellites serving 109 member nations. INTELSAT satellites over the Pacific Ocean provide service between the continental United States and Hawaii, Japan, the Philippines, and Thailand. Other INTELSAT satellites provide service across the Atlantic. COMSAT is a managing agency for INTELSAT.

Intersputnik is the outgrowth of the Intersputnik Agreement concluded in Moscow in November 1971. It is really a major communications satellite network operated by the USSR and serving the USSR, Eastern Europe, and other Soviet bloc countries.

ESA was adopted in May 1975 and came into force in October 1980. Designed to replace ESRO (European Space Research Organization, 1962) and ELDO (European Launcher Development Organization, 1964), it is a thirteen-member consortium of European nations whose objective is to promote, for peaceful purposes, cooperation among European countries in space research and technology and their space applications. ESA is heavily engaged in building and launching communications satellites, experiments in MPS, building a mini space shuttle called *Hermes*, and launching communications satellites via its unmanned *Ariane* rockets. Arianespace, the commercial arm of ESA, provides space insurance and other services for ESA countries.

ARABSAT was established by the members of the Arab League in 1976. Its purpose is to use an Arab satellite (one is in orbit) as a means of serving the purposes of communications, information, culture, and education in the Middle East. ARABSAT is structurally similar to the other regional agencies, having the member countries' ministers of communications as the supreme body, a board of nine directors, and a general manager. Financial contributions to operate the system are made by member countries, which obtain a proportionate share of the profits.

These regional agencies have some notable common features. They avoid the legal tightness of some organizations and, though similar in internal structure to UN specialized agencies, they preserve a pragmatic flexibility and are in practice partnerships or clubs. They are not wholly intergovernmental in function, for their operations must depend on cooperation with foreign national agencies and private enterprises through service and supply contracts.

OTHER SPACE OPERATORS

This chapter would be incomplete if it did not mention astronauts and cosmonauts as true space operators. History is replete with examples of these daring space operators. In 1961, astronauts from the United States and USSR went into Earth orbit for the first time. The United States conducted a series of manned space flights in *Gemini* spacecraft in 1965–1966 and eleven flights in *Apollo* spacecraft from 1968 to 1972. U.S. astronaut Neil Armstrong was the first human to walk on the Moon, in July 1969. In 1973, astronauts spent thirty, sixty, and eighty-four days in orbit in three *Skylab* operations. There have been several manned flights of U.S. space shuttles, and Soviet cosmonauts have spent several months at a time in space aboard their *Salyut* space stations.

As a result of these manned space flights, a great deal has been learned about the difficulties of such flights and their practical solutions. The effects of high acceleration and weightlessness in conditions of microgravity can be tolerated, though the latter may cause space sickness; and temperature in the spacecraft can be controlled, particularly the overheating of the sunlit side of the spacecraft. But radiation in the Van Allen belts, in cosmic rays, and in high-energy particles in occasional solar flares can be a constant hazard, and exposure must be prevented or limited by metallic shielding or the creation of magnetic fields in the spacecraft.

The skin of the spacecraft is vulnerable to meteoroids or dust particles. The requirement of a reduced weightload for effective orbiting restricts the volume of space for occupancy and also the quantity of materials that may be carried, particularly food. It may be that the development of hydroponics[1] or making edible food from the activity of bacteria will resolve part of this food problem. Another solution is carrying lightweight and compact food on space journeys. But it seems that it will be a long time before space travel can be much extended in time or numbers of travelers.

The United States and USSR have separate plans to send manned flights to Mars, and there is some talk at NASA about establishing small colonies of around thirty persons on Mars within the next fifty years (*Pioneering the Space Frontier* 1986). Residence on Mars or the Moon is still a remote possibility. In pursuit of this objective, obvious needs include more powerful spacecraft, more space suits for protection from radiation and for self-supporting breathing apparatus, and elaborate housing for protection and for food production.

COMPETITION AND COOPERATION

There is a great deal of competition among space operators for the economic spoils of outer space. The superpowers—the United States and the USSR—are competing for leadership and technological superiority in the space race. They are emphasizing satellite communications, MMPS, remote sensing, and strategic military uses of outer space. The FRG, other members of ESA, the PRC, and Japan are pursuing peaceful uses of space, such as MMPS, remote sensing, communications technology, and space transportation. NASA and ESA are serious rivals in the space transportation and satellite launching business, and the PRC has entered that competition (Gengtao 1987; Covault 1984a). Developing countries, such as India, Pakistan, and Brazil, are too poor to be considered serious contenders in the space race. Private companies in all these countries are competing for lucrative contracts from their respective space agencies.

There is also a great deal of cooperation among countries (and among companies) in space business. In the thirty years from 1957 to 1987, the United States has signed more than 1,000 agreements with 100 countries for international space activities (Covault 1987a). The U.S./International Space Station is a collaborative project with ESA, Japan, and Canada; and the FRG, the UK, and Israel are assisting on SDI projects. The Soviet Union's *Mir* space station has Soviet bloc and French participation. The exploration of Mars offers an opportunity for future U.S.-Soviet cooperation.

SUMMARY

This chapter discusses the primary operators in space—private entrepreneurs, government agencies, and international agencies—and their working relationships. It also mentions another important group of space operators: astronauts and cosmonauts. Competition and cooperation among countries, government space agencies, and companies for space business are also examined.

Space business generally requires huge amounts of capital and long-term commitment. Corporations are driven by profits; government agencies by prestige, glory, and international reputation; countries by economic development, profits, and technological pride; and astronauts by personal achievement and national pride. Governments create the laws and regulations under which the private sector operates. In space business, many of these regulations are fuzzy or evolving.

NOTE

1. Hydroponics is the cultivation of plants using, in place of soil, solutions of the mineral salts that the soil normally provides.

CHAPTER 4

Materials Processing in Space

Space offers a combination of physical properties almost impossible to duplicate on Earth: the absence of vibration, lack of convection, unlimited heating and cooling, lack of atmospheric attenuation, near-perfect vacuum, unfiltered sunlight, and sterile environment. Most important, space (or rather orbit) gives scientists a place to work without interference from gravity, often referred to as microgravity or zero-gravity (zero-g).

The gravitational attraction of Earth on a spacecraft in a 400-kilometer orbit is only about 12 percent less than it would be if it were on the Earth's surface. The phenomenon of weightlessness occurs because the spacecraft is in a state of free fall. A spacecraft that has achieved orbital velocity has a gravitational environment of about 10^{-7} g or one ten-millionth of Earth's gravity. The situation existing in orbit is referred to as a microgravity environment.

Microgravity allows certain enzymes to separate more readily than in a gravity-influenced environment or some chemical elements to mix that cannot be mixed in gravity on Earth. Hence, the attraction of materials processing in space for industry is the opportunity to produce products in space that cannot be produced economically on Earth or in sufficient quantities. It is expected that space-made products will have greater purity, strength, quality, and uniformity.

ADVANTAGES OF MICROGRAVITY

The microgravity environment opens a new frontier for materials scientists. It will provide new knowledge about the effect of gravity on material properties and process mechanisms, and it will eliminate some of the disruptive effects of gravity that prevent some Earth-bound materials from achieving theoretical performance characteristics. Microgravity eliminates convection, sedimentation and buoyancy, and hydrostatic pressure, and it allows containerless processing.

Convection

Gravity-driven convection—stirring of mixtures of liquids or gases with slightly nonuniform densities—can cause undesirable stirring and mixing during the growth of crystals, the casting or solidification of alloys and composites, the reaction processes of chemicals, and the separation of biological materials. Thus, its elimination can benefit a number of materials processes:

- Producing semiconductors with more uniform composition for use as chips in advanced computers, defense systems, and other electronic hardware.
- Preventing unwanted mixing of electrokinetic separation techniques, such as those used to purify pharmaceutical materials.
- Studying the mechanisms responsible for macrosegregation and microsegregation in castings to determine how to make more uniform castings on Earth with fewer defects that are stronger.
- Studying fluid dynamics in microgravity to further the science of fluid dynamics and to identify new applications and chemical processes.

Sedimentation and Buoyancy

The elimination of gravity-induced sedimentation and buoyancy (high-density materials settle and low-density materials float in a liquid or gas mixture) can create new alloys and composites by permitting particles of vastly different density to remain in uniform suspension until solidification. Also eliminated is the need for mechanical stirring, important since stirring may be detrimental to the materials involved. The resulting beneficial applications include these:

- New alloys that are immiscible (cannot be mixed homogeneously) on Earth.
- More uniform or homogeneous composite materials and alloys from constituents with large density differences.
- Large semiconductor and electro-optical single crystals (because pieces do not break away during the crystal growth process).

Hydrostatic Pressure

The lack of hydrostatic pressure (gravity-induced pressure exerted by liquids at rest) under microgravity conditions means that liquids or solids do not deform under their own weight. The resulting beneficial applications include the growth of large, single semiconductor and electro-optical crystals without deformation and with fewer defects, and the fabrication of intricate castings, such as directly solidified turbine blades, without deformation by putting each finished machined part in a thin ceramic skin and then melting and resolidifying it in microgravity.

Containerless Processing

Microgravity offers new processing possibilities because it eliminates the need to confine liquids and molten materials within a container. Using acoustic, electromagnetic, or electrostatic fields, it is possible to mix, manipulate and shape, and solidify liquids in free suspension. Surface tension holds the materials together in a mass. Containerless processing eliminates wall-induced contamination, solidification nucleation, and strain, thereby improving the optical or physical properties of a material.

Advantages of Infinite Vacuum

The high vacuum of space could be an important resource for the ultrahigh purification of materials in space. Several purification techniques—such as vacuum melting, vacuum distillation, zone refining, and electrotransport—require an ultrahigh vacuum during processing; all need high heat loads and large gas loads. Use of these processes on Earth is difficult and costly because the large vacuum systems required are generally too costly for high-volume applications. The free vacuum of space may not only provide technical advantages but may be more economical as well. Used with containerless processing, these techniques should improve the purity of materials enormously.

COUNTRIES AND ORGANIZATIONS INVOLVED IN MPS

The leading countries involved in MPS experiments are the United States, USSR, ESA member countries, and Japan. Each is experimenting with pilot production of pharmaceuticals in space (interferon, Factor VIII, urokinase); unique metal alloys; special plastics; special crystals for high-speed computers and other electronic and defense applications; ultrapure glass for telescopes, cameras, and other equipment; and stronger ceramics, to name a few. Electrophoresis in space could produce purer and higher-quality pharmaceuticals than on Earth. Made-in-space metal alloys could be lighter but ten times stronger than those made on Earth. Pure and perfect gallium arsenide crystals (made from the two soft metals gallium and arsenic), which conduct electrons ten times faster than silicon, can be used in computer chips, lasers, switching devices in fiber-optic systems, high-frequency antennas, solar-power arrays, and other electronic devices (Osborne 1985, p. 47). They are particularly useful in anything sent into space because of their resistance to radiation and heat and their tremendous performance speed. When grown on Earth, these crystals suffer impurities, striations, and lack of uniformity, which reduces and adversely affects their performance. Ultrapure glass in telescopes and cameras will allow us to see clearer and farther into the universe.

In the United States, some of the leading companies conducting MPS experiments are 3M (crystals, plastics, glass), IBM (crystals), John Deere (metal alloys), General Motors (metal alloys, lubricants), Aluminum Corporation of America (ALCOA—aluminum), Grumman Aerospace (metal alloys), Union Carbide (glass-forming alloy systems), Westinghouse (ultrapure, bubble-face glass), McDonnell Douglas (pharmaceuticals), and Microgravity Research Associates (MRA) of Coral Gables, Florida (crystals). Some of the leading U.S. universities engaged in some aspect of MPS research are the University of Colorado at Boulder (Research Center for Low-Gravity Fluid Mechanics and Transport Phenomena) (Johnson 1986, p. 66), the University of Houston, and the Ohio State University. The Marshall Space Flight Center, in Huntsville, Alabama, is the primary NASA location engaged in MPS research (Dooling 1985, p. 14).

While the United States has run between 100 and 300 MPS experiments, the Soviets have conducted some 1,500. They have developed infrared receivers using cadmium mercury telluride crystals made in space; they have tested the synthesis of interferon, urokinase, and other drugs; and they have produced metal alloys and special glasses for optical devices. West Germany, France, and Japan are building space platforms on which to conduct MPS in the future and are gearing up to do MPS once the U.S./International Space Station becomes operational. Major space agencies involved in MPS are NASA, NASDA in Japan, ESA, the USSR's Glavkosmos (counterpart to NASA), and the British National Space Centre.

POTENTIAL PRODUCTS AND MARKETS

MPS is not just building factories in space. There are three aspects to it: R&D to improve terrestrial processes, R&D leading to space production for later terrestrial sales, and technology transfer (Dooling 1985, p. 16).

First Commercial Made-in-Space Product

The first commercially successful made-in-space products were latex microspheres (sometimes called latex beads) produced on a U.S. space shuttle flight in 1983. Microspheres are used as calibration targets (devices) for electron microscopes, pollution monitors, paint and toner production, and medical diagnostic work. The upper limit on the size that can be made on

Earth is 3 microns; larger sizes can result in creaming and coagulation because Brownian motion no longer keeps them suspended in the solution. Weightlessness provides that suspension (Dooling, 1985, p. 16).

The National Bureau of Standards (NBS) at Gaithersburg, Maryland, sold the spheres in 600 5-milliliter vials of 30 million spheres for $384—or $434,000 an ounce (Dooling 1985, p. 16). NBS estimates that around $3 million worth of microspheres can be produced per U.S. space shuttle flight, and it would like spheres as large as 100 microns. Despite the program's success, follow-on flights have been postponed indefinitely, especially in the aftermath of the *Challenger* tragedy. This has caused an erosion and loss of skills as members of the science team become frustrated and move on to other projects.

Pharmaceuticals

The pharmaceutical industry is likely to be the first user of space for manufacturing; the weightless environment is ideal for the delicate separation of complex, nearly identical substances. Jastrow (1984) states that some rare and expensive medicines could be made more purely, cheaply, and efficiently in space than on Earth using electrophoresis. These medicines include urokinase, Factor VIII, and the beta cell.

The Center for Space Policy, a consulting firm in Cambridge, Massachusetts, forecasts that if given suitable environment for development, by the year 2000 space industries in the United States could be producing $27 billion annually in pharmaceuticals to battle cancer, emphysema, diabetes, and other diseases (Center for Space Policy 1985).

Scientists foresee the manufacture of these pharmaceutical products on space stations and free-flying (automated) space platforms with compartments and laboratories leased (or purchased) by industry. Manufacturing would be done by robots and workers in space. Basic research would be performed by scientists who would work on the U.S./International Space Station for about three months at a time. Before they returned to Earth, their places would be taken and their work continued by new crews. Shuttles would pick up both the scientists and the processed products and return them to Earth.

The McDonnell Douglas Corporation is the leading U.S. corporation engaged in experiments to produce pharmaceuticals in space. It has done work on a number of substances, including urokinase and interferon (a cancer drug that blocks or inhibits the growth of cancer cells) and is experimenting with the production in space of erythropoietin, the kidney hormone that stimulates red blood cell production (and so helps to treat or prevent anemia, among other conditions). The company estimates that 2 million people, about 1 percent of the U.S. population, are anemic and that 600,000 to 800,000 people per year might benefit from the drug (Byars

1985e). It is estimated that McDonnell Douglas has invested more than $20 million to date in pharmaceutical MPS experiments, equipment, and facilities.

In its market analysis before embarking on its electrophoresis operations in space program, McDonnell Douglas investigated biological products amenable to electrophoresis (such as hormones, enzymes, cells, and proteins) and identified twelve products suitable for profitable production in space (Table 4.1). Potential U.S. annual sales for these products are estimated to be between $2 billion and $7 billion, depending on the proportion of patients with each condition that use them. The world market for these products (including the United States, Western Europe, and Japan) could be about $20 billion annually.

Electronic Crystals (Semiconductors and Electro-Optical Materials)

Primarily because of the greater purity and reduced defects achievable in space, microprocessor chips made from space-produced electronic crystals are expected to perform up to ten times better than their Earth-produced counterparts. Thus, given the rapid growth and intense competition of the semiconductor, computer, and information-processing industries, space-grown crystals could produce highly competitive semiconductor and electro-optical devices. Like pharmaceuticals, electronic crystals are high-value materials amenable to production in small, automated facilities in orbit.

The three electronic materials with the greatest potential for production in microgravity are gallium arsenide for microwave and high-speed electronic signal-processing circuits, cadmium telluride substrates for mercury cadmium telluride infrared detector arrays, and indium phosphide for near-infrared optical devices and high-speed signal-processing circuits. The improved quality of these electronic materials could lead to many new military and commercial applications.

A 1983 survey of the U.S. semiconductor industry and U.S. government semiconductor users conducted by Rockwell International with Kern and Associates found that gallium arsenide crystals of higher quality than are available are needed for very high-speed microwave circuits, radiation-resistant high-speed signal processing on missiles, high-speed signal processing with integral lasers for readout through fiber optics, and semiconductor radar arrays on planes and satellites. The impurities and crystal dislocations of gallium arsenide crystals made on Earth are too high for these applications.

Some experts believe that these high-performance applications could justify material costs up to 100 times the cost of Earth-grown materials. By 1995, sales of these space-grade crystals could be $65 million ($25 million for gallium arsenide and $20 million each for cadmium telluride and indium

Table 4.1
Candidate Pharmaceutical Products for MPS

Typical Products	Treatment Target	Status/Potential	Estimated Annual U.S. Patients (Thousands)
a_1 Antitrypsin	Emphysema	High purity possible in space; only research quantities available on Earth	100
Antihemophilic Factors VIII and IX	Hemophilia	High yield possible in space	20
Beta cells	Diabetes	Can separate commercial quantities in space but not on Earth; potential for single-dose cure	600
Epidermal growth factors	Burns	High purity possible in space; potential for replacement skin grafting	150
Erythropoietin	Anemia	High purity possible in space; potential for replacement transplants and transfusions	1,600
Immune serum	Viral Infections	Higher purity possible in space	185
Interferons	Viral Infections	High yield and high purity possible in space; possible unlimited potential	10,000
Granuloctye stimulating factor	Wounds	Only research quantities now available on Earth	2,000
Lymphocytes	Antibody production	Potential to replace antibiotics and chemotherapy	600
Pituitary cells	Dwarfism	High purity possible in space speculative treatment	850
Transfer factor	Leprosy/multiple sclerosis	Potential for several applications	550
Urokinase	Blood clots	Low development costs possible	1,000

phosphide) according to the market study. Estimated prices for the crystals are $600 per gram for gallium arsenide between 1988 and 1992 and $250 per gram thereafter; $2,000 per gram for cadmium telluride between 1988 and 1994 and $1,000 per gram thereafter; and $1,000 per gram for indium phosphide between 1988 and 1995 and $500 per gram thereafter. The drop in prices over time assumes higher-volume production and the existence of a permanent space station or space platform that will reduce production and transportation costs. These dates will have to be pushed back at least ten years in the aftermath of the *Challenger* accident.

Other Products

Among the many other products being considered for experimentation and production in space are metal alloys, glass, and ceramics. Both government and private industry hope to develop more than 500 alloys that can be made only in zero-g (Miglicco 1985, p. 36). One such new metal would give jet turbine blades increased resistance to high temperatures while maintaining their strength and light weight. The development of this type of blade is essential to the next generation of jet engines.

Ultrapure, bubble-free glass can be made by using a containerless and vibration-free manufacturing process. Such glass could improve laser technology, which in turn would advance the communications industry and further work in fusion reactors. Fiber optics also could be improved because such products would significantly outperform their counterparts made on Earth (Miglicco 1985, p. 37). Space-based techniques, including containerless processing made possible by microgravity, may allow the production of high-strength, lightweight, heat-resistant ceramics.

MICROGRAVITY FACILITIES FOR RESEARCH AND PRODUCTION

There are basically three categories of facilities for research production relative to MPS: facilities on Earth, the U.S. space shuttle, and future spacecraft such as space stations and space platforms.

Facilities on Earth

Before committing themselves to shuttle-launched experiments, companies can perform preliminary experiments in ground facilities such as the 340-foot drop tube facility at the NASA Lewis Research Center in Cleveland, Ohio, or the drop tower facility at the NASA Marshall Flight Center in Huntsville, Alabama. Both provide 4.6 seconds of low gravity for small samples during free fall. Aircraft and rockets that provide low-gravity conditions for 15 to 60 seconds and 4 to 6 minutes, respectively, during the peak of parabolic trajectory are also possible sites for experimentation. A few U.S. companies, such as McDonnell Douglas and 3M, are looking into the possibility of developing ground-based facilities that duplicate the microgravity environment of outer space.

U.S. Space Shuttle

The space shuttle also supports inexpensive proprietary experimentation through "Getaway Specials" in small, self-contained payload canisters. The canisters come in three sizes: 2.5 cubic feet (0.07 cubic meters) weighing a maximum of 60 pounds (27 kilograms) for $3,000; 2.5 cubic feet (0.07 cubic meters) at no more than 100 pounds (45 kilograms) for $5,000; and 5 cubic feet (0.14 cubic meters) weighing a maximum of 200 pounds (91 kilograms) for $10,000. Because Getaway Specials are self-contained, they must operate automatically (except for three on-off switches to allow operations by the shuttle crew), and each unit must have its own power supply.

A space shuttle, however, is not an ideal environment for extensive MPS, for at least three reasons. First, requirements for MPS emphasize high power and low gravity. A g-level of less than .00001 g is typically required continuously for several weeks. Manned systems, such as the space shuttle or overly large structures such as a space station, would tend to vary from such g-limits frequently due to control activity, crew motions, gravity gradient effects, or even molecular air resistance. Second, total mission duration is rarely longer than seven to ten days—too short a time for meaningful and substantial MPS. Third, space for MPS payloads or experiments on the shuttle is limited.

Other Spacecraft

Other spacecraft being developed for MPS and other orbital activities are space platforms, such as ESA's EURECA; the shuttle pallet satellites (SPAS) being built in West Germany; and the ISF being developed by Space Industry, Inc., Houston, Texas. Space platforms are also being designed separately by the Soviets and the Japanese.

The Soviets are conducting MPS activities on their newest space station, *Mir*, the core of which was launched in 1986. The United States, ESA, Japan, and Canada are collaborating on the development of the U.S./International Space Station, which will be used partly for MPS or MMPS.

OBSTACLES TO MPS

MPS faces a number of severe problems. This is not to suggest that commercially viable MPS will never occur; it will, but many decades from now, and for a few select products.

Transportation

Transportation into space is essential for the development of MPS. The present lack of regular, certain, and on-time transportation schedules into space—especially since the *Challenger* tragedy—has all but crippled the MPS industry in the United States. It is estimated that NASA's backlog of MPS experiments numbered around 450 by the end of 1986 (Vaucher and Robertson 1986, p. 46). Many firms known for their MPS experimentation, such as McDonnell Douglas and John Deere, have suspended their experiments indefinitely and redirected some of their research to terrestrial experiments. Others have decided not to pursue space ventures further. The job of persuading management of the value of continuing to invest in space research programs, never easy, has become considerably more difficult. Even when the shuttle is available, manifesting and integration difficulties are such that customers have to wait three to five years to fly—hardly a way for firms to do business. And if products are made in space, some may not survive the ride back to Earth.

MPS Costs

Space-based manufacturing is very costly: $50 million and more. This cost amount includes plant construction in space, factories or laboratories for transport into space, equipment, salaries for high-tech scientists and engineers, interest expense on huge sums of borrowed capital, and transportation into space (about $10,000 per kilogram) (Shifrin 1984). Prices for manufactured products in space will have to be between $100,000 and $1 million per kilogram (*Economist* 1984a, 1984b). As long as these costs remain high, MPS will grow very slowly because only a few wealthy firms can afford these costs, and even for them, product prices reflecting these costs may be too high to attract sufficient economic demand on Earth.

Economic Demand and Markets

The absence of already developed markets—or at least the uncertainty about future markets—is probably the biggest obstacle to MPS. Even if some pharmaceuticals can be made purer and better in the microgravity environment of space, will the added cost of making them in space increase their price to such an extent that they are priced out of the market? Will pharmaceuticals capable of being effectively produced in space find a large enough market on Earth? In the United States alone, it is estimated that cancer killed 483,000 people in 1987 and that there were 965,000 incidences of cancer in the United States that same year (American Cancer Society 1988). In 1986, 36,340 people died from diabetes, and 6,585,000 had chronic diabetes. The incidence of diabetes per 1,000 in the United States is 28 (National Center for Health Services 1988). Are these markets large enough to justify pharmaceutical production in space? Maybe, if one thinks of the worldwide market, or the recurring nature (repeat business) of the market.

Some believe that Earthbound biotechnology techniques do just as well. In early 1985, Johnson & Johnson and McDonnell Douglas were collaborating on the production of erythropoietin (medicine for stimulating red blood cell production to combat anemia and other disorders) and other medicines. By mid-1985 Johnson & Johnson pulled out of the effort because of a belief that Earthbound processing techniques could provide similar results more quickly and less expensively (Byars 1985e; *Aviation Week & Space Technology* 1985m).

For some products, such as metal alloys, MPS does not appear to be prudent. The metal industry is not expected to have a sizable market for space processing, at least not for fifty years, because the value of the material is too low to justify MPS. Research in space will, however, provide new insights into materials science on Earth and may allow some firms to process sample quantities of special lightweight, high-strength alloys, but commercial revenues are not expected to be significant.

Knowledge Base

Firms involved in MPS are still basically unfamiliar with MPS requirements. It will take several years of research for them to build up a knowledge base sufficient to move from the experimental stage in MPS to large-scale, commercially viable MPS.

Profit Squeeze

Given the huge capital outlay required for space ventures, profits will generally be nonexistent for several years after initial investments. Payback periods may be ten years and beyond, with uncertainty about future profits.

Technology

Some of the foremost barriers in the commercial development of space are technological. Generally these are caused more by unfamiliarity with hardware design for space use than by the difficulty in designing the hardware. These technological barriers will retard the development of MPS and other orbital operations. Furthermore, technological facilities are often very expensive ($10 million and more) and take several years to develop.

Space Insurance

Insurance coverage is a facilitating marketing factor (Kotler 1988). In the uncertainty of and potential dangers posed by the space environment, insurance coverage for workers in space, expensive space-based assets like space stations and space platforms, and third-party liability assumes a role of paramount importance. In short, the availability of space insurance at reasonable premiums is essential for the growth and success of MPS and other space-based activities.

But obtaining this sort of insurance coverage is very difficult and costly due to the failure of several shuttle and *Ariane*-launched satellites during 1984–1986 and the *Challenger* tragedy. These failures and accidents led to more than $600 million in losses to insurance underwriters. Those that remained in the space insurance business increased premiums for space ventures from 10 percent to 30 percent of insured value and developed stricter policies (*Aviation Week & Space Technology* 1985k, 1985l, 1985p, 1985r; Payne 1985).

Insurance problems also create marketing problems. In 1985, Fairchild Industries (U.S.) cancelled its $200 million leasecraft program to build a space platform to provide in-orbit electrical and support services for a range of MMPS customers. This cancellation resulted when the expected customer base did not emerge, in part because Fairchild could not guarantee insurance coverage to customers (Fink 1985). As long as space insurance is very costly and difficult to obtain, MMPS or MPS will be very limited.

Health Problems

If space-based manufacturing of any kind is to succeed, the pharmaceutical industry and medical scientists will have to do extensive research aimed at solving health-related problems of humans working and living in space. These problems include bone decalcification/deterioration in space brought on by inactivity under weightless conditions and the fact that calcium leaches out of the bone in space through urine; loss of muscle mass and red blood cells; impairment of the immune response; and problems of cardiovascular deconditioning after months of exposure to microgravity (Chase 1986). There are doubts that humans can work effectively and efficiently in weightlessness longer than three months or so at a time.

Soviet cosmonauts who returned to Earth in 1984 after spending 237 days in space emerged from the flight with symptoms that mimicked severe cerebellar disease or cerebellar atrophy. (The cerebellum is the part of the brain that coordinates and smooths out muscle movement and helps create the proper muscle force for the movement intended.) These cosmonauts required forty-five days of Earth gravity before muscle coordination allowed them to remaster simple children's games, such as playing catch or tossing a ring at a vertical peg (*Pioneering The Space Frontier 1986*, p. 128). They also had to relearn how to walk.

Space-based medical research by the USSR and West Germany also indicates that microgravity may have a stimulating effect on the growth of bacteria in space and may render bacteria in space more resistant to antibiotics (*Commercial Space* 1986). These results suggest a high risk of acquiring new diseases in space and pose serious questions about the human ability to ward off diseases while living in space for extended periods of time.

Federal Regulations

Federal regulations represent other obstacles to MPS. These regulations may pertain to a host of U.S. federal agencies, such as the Food and Drug Administration (FDA), the Federal Communications Commission (FCC), the Department of Defense (DOD), and the Department of Transportation (DOT). The FDA will be used here as an illustrative example.

Inspections of Space Factories. Inspection of factories in which drugs are manufactured is allowed by the Food, Drug and Cosmetic Act (FD&C Act) executed by the FDA. Inspections in outer space will be difficult, if not impossible. Further, lack of inspections may make it difficult for the FDA to ensure that good manufacturing practices (GMPs) are being followed. One could argue, however, that the FDA could simply concern itself with the finished product and not with factory inspections. Yet failure to follow current GMPs means that the finished product is adulterated and in violation of the FD&C Act, though the finished dosage may be all it is supposed to be. If the FDA's ability to confirm the use of current GMPs is impaired, the product may be found adulterated despite the purity and efficacy of the product returned to Earth; and disapproval is likely. Other inadequacies of present regulations, when applied to outer space, will become apparent when manufacturing actually begins.

New Regulations. Drugs produced in space, although similar in nature to the equivalently titled drug produced on Earth, will undergo substantial change in composition and purity and will therefore be subject to new drug regulations under the FD&C Act. Further, because of the newness of the technology and the unpredictability of factors affecting it (e.g., cosmic radiation), investigational studies requiring evaluation through applicable investigational drug regulations will be necessary before premarket clearance is considered. Drugs produced in space may consequently be subject to a potpourri of regulations. The pace and direction of drug production in space will be influenced significantly by the form and complexity of these regulations, covering product standards, premarket approval, disclosure to users, and other areas. The extent to which FDA regulations will affect MPS operations is not known. What is certain, however, is that new FDA regulations relevant to space-based manufacturing will be necessary, time-consuming, and costly. They may also act to inhibit pharmaceutical production in space.

FDA Authority over MPS and the Agency's Relationships with Other Federal Agencies. Food and drug law deals with a government's attempt to protect public health and individual welfare in the development and marketing of essential goods. Over the years, regulation has reflected concern for purity, safety, and the efficacy of drugs. But the U.S. Congress has never contemplated the FDA's regulation of pharmaceutical production in space, and the FDA has no statutory indication of its regulatory jurisdiction in the area or whether it conflicts with the authority of other federal agencies. These issues will have to be addressed if and when commercial production in space begins.

As long as drugs produced in space are manufactured for the purpose of placing them in interstate commerce, they are subject to FDA regulation. The agency's broad consumer protection mandate has, however, brought it into occasional conflict with other federal agencies, thereby making it necessary to establish a cooperative effort between the FDA and other agencies, such as the Federal Trade Commission, the Consumer Product Safety Commission, the Environmental Protection Agency, the U.S. Department of Agriculture, and the Occupational Safety and Health Administration. This cooperative effort serves to avoid conflict and to improve efficiency of regulatory effort.

Dealing with NASA. The advent of MPS will involve the FDA with federal agencies it has not previously had to deal with, such as NASA. Civilian space activities in the United States are under the exclusive direction of NASA, so although McDonnell Douglas and pharmaceutical companies are concerned about FDA approval of drug applications coming out of MPS, they have not (at least not publicly) consulted the FDA about possible MPS procedures. Apparently the companies see NASA as the agency primarily responsible for approval of the project. But when production is fully underway and the FDA processes new drug applications, the FDA will be concerned about GMPs, drug safety, purity, quality, efficacy, and potency. Initial FDA involvement will be appropriate to, if not necessary for, a smoother entry of space-made pharmaceuticals into the marketplace.

NASA approval of the testing methods and procedures used in MPS does not guarantee approval by the FDA. The FDA has no access to the NASA projects concerning pharmaceutical production in space and no working relationship with NASA. This is so, perhaps, because MPS is in its infancy or, more correctly, in its embryonic stage. But as MPS grows, the FDA and NASA will have to work more closely together and coordinate their work with other agencies.

Age of Substitutability

A final economic fact of life that is adverse to space-based manufacturing is the development of Earth-made products that are good substitutes for space-made products, thus reducing the importance of or reliance on MPS. For example, scientists are learning to replace rarer, specialty metals with high-performance ceramics, plastics, and glasses made from the Earth's abundant supply of clay, silicon, aluminum, hydrocarbons, and iron.

SUMMARY

This chapter discusses various aspects of MPS, such as the advantages of microgravity for MPS, potential MPS products and markets, microgravity facilities for research and production, and impediments to MPS or MMPS. The advantages of microgravity for space-based manufacturing are that it eliminates the phenomena of convection, sedimentation and buoyancy, and hydrostatic pressure, and allows containerless processing. These advantages will tend to give rise to space-made products that are generally more uniform in composition and purer, stronger, and lighter than their Earth-made counterparts. Additionally, microgravity allows for much higher yields—estimated as much as five to seven times more—for space-made products using electrophoresis.

Some of the potential products being considered for MPS are pharmaceuticals like interferon, urokinase, Factor VIII, and the beta cell; metal alloys; specialty ultrapure, bubble-free glasses for telescopes, cameras, and other devices; perfect crystals for powerful high-speed computers and other electronic markets; and ceramics. Deudney (1982) and others believe that some pharmaceuticals and alloys are the most likely products to be made in space. Like many other scientists, however, Deudney is not optimistic about commercial MPS or MMPS for a variety of reasons, including the tremendous obstacles to space-based production, such as costs, lack of present economic demand for space-made products, lack of on-time and reliable transportation schedules into space, difficulties in obtaining space insurance, the need to develop space factories, and the ever-present possibility of developing equally good substitutes on Earth much less expensively than space-made counterparts.

MPS will continue to be done by large, wealthy companies like McDonnell Douglas and 3M, which envisage the microgravity environment of space as advantageous for producing pharmaceuticals and special alloys that are superior to made-on-Earth alternatives. Such companies also hope that the materials science lessons learned through MPS will be useful for terrestrial applications. Because of the present obstacles to MPS, commercially viable MMPS is several decades away, beyond the year 2000.

CHAPTER 5

Spin-offs and Market Opportunities

SPIN-OFFS IN VARIOUS FIELDS

Public Safety

Aerospace technology has been beneficially transferred to such civil applications as protective undergarments for workers in hazardous environments, fire-retardant paints and foams, fireblocking ablative coatings of outdoor structures, flame-resistant fabrics for use in homes, offices, and public transportation vehicles, and a NASA-developed breathing system for firefighters.

In the 1960s, the Celanese Corporation of New York developed for NASA and the Air Force Materials Laboratory a fireblocking fiber known to chemists as polybenzimidazole (PBI). PBI fiber emits very little smoke or "offgassing" at temperatures up to 1040°F. Fabrics made from PBI are durable and comfortable, do not burn in air or melt, have very low shrinkage at high temperatures, and are resistant to strong acids, solvents, fuels, and oils. PBI was used in *Apollo* and space shuttle astronaut gear and in webbings and tethers on *Apollo* and *Skylab* flights.

Beginning in the 1980s, PBI found many new civil applications as markets opened for an alternative material to asbestos and in response to stricter government antipollution standards. Some of these new commercial applications are thermal protective wear for foundry workers, chemical plant employees, firefighters, auto racing drivers, and others whose occupational activities expose them to flame and intense heat; and a range of PBI airline seat fabrics covering the foam seat cushion.

Celanese's main PBI-producing plant is located at Rock Hill, South Carolina. It has expanded its market for PBI fibers into the Far East and Europe through agreements with Teijin Limited, Osaka, Japan, and Hoechst A.G., Frankfurt, Germany.

Another public safety spin-off from space technology is the breathing apparatus worn by firefighters for protection from smoke inhalation. The apparatus includes a face mask, a frame and harness, and a basic air cylinder with its associated warning device, valves, and regulators. The traditional breathing system was heavy, cumbersome, mobility restricting, and so physically taxing that it induced extreme fatigue.

The Johnson Space Center (JSC), in collaboration with such companies as Martin Marietta, Structural Composites Industries, and Scott Aviation, developed a breathing apparatus weighing slightly more than 20 pounds (about one-third less than predecessor systems), a face mask that offered better visibility and fit, and with the air depletion warning device designed so that the beeping alarm could be heard only by the wearer, thus minimizing confusion in the hectic environment of a fire. The breathing systems emanated largely from the portable life support systems used by Apollo astronauts. As a result of the lightweight breathing systems, inhalation injuries and fatigue to firefighters have been drastically reduced.

Health and Medicine

There are several space technology spin-offs into medicine. For example, in the early 1970s, NASA's Jet Propulsion Laboratory (JPL), Pasadena, California, undertook a community service project to meet a need of the Los Angeles Police Department. The department's forensic chemistry laboratory wanted a reliable, rapid way to detect drugs in blood or urine samples taken from suspected narcotics users. JPL invented a technique for speedier separation of biological compounds and developed an automated system for analyzing the extracted compounds. The automated system is called AUDRI (automated drug identification). NASA granted licenses for commercial use of the technology to three companies; one of them is Analytichem International, Harbor City, California, whose products have found wide and growing acceptance all over the world in police work, the pharmaceutical industry, and the chemical industry.

Digital imaging is another space technology spin-off that has found its way into medicine (and industry). Image analysis is the art of obtaining information from pictures—for example, through visual examination of photographs or X-rays. But visual extraction and interpretation of information is slow, tedious, and error prone because it is subjective. To support space requirements, NASA—in particular JPL—developed the technique of digital imaging (computer-processed numerical representation of physical images, such as of the planets and moons of the Solar System). JPL also played a lead role in improving the quality of images and making them easier to interpret.

Today there are scores of nonaerospace applications of digital imaging. In medicine, CAT scanners and diagnostic radiography systems are based on digital imaging; three-dimensional reconstruction techniques are proving a valuable aid to microscopy; and computerized image analysis of cardiological X-rays is providing quantitative data on heart valve and artery functions. In industry, high-resolution digital imaging systems are employed in quality control inspection systems, and there are other applications in chemistry, cartography, manufacture of printed circuits, metallurgy, ultrasonics, and seismography.

Another spin-off of space technology in medicine is the scalp cooler. The scalp cooler, introduced to the market in 1985, consists of a fabric head cover with a network of flexible plastic tubing through which a coolant, usually cold water, is pump circulated from a container adjacent to the patient. The scalp cooler has electronic components to show scalp temperature and to set the required temperature range. The basic technology stems from JSC's development of a liquid cooling undergarment, worn beneath a space suit, through which coolant is circulated to remove the excess body heat of astronauts.

The scalp cooling system, known as CHEMO-COOLER Treatment Support System, is produced by Composite Consultation Concepts, Inc. (CCC), Houston, Texas. It is a method of combating hair loss in cancer patients undergoing chemotherapy. It has been found that lowering the scalp temperature reduces the amount of drug absorbed by hair follicles and prevents hair loss in many patients yet does not weaken the drug's anticancer effect.

Transportation

In the field of transportation, spin-offs from space technology have been abundant. One is the development at NASA of computer software to aid in the improved design of automobiles and aircraft. Industry has improved on software and modified it to suit specific needs. An example is SPAR (structural performance and design), a computer program developed by NASA's Lewis Research Center, Cleveland, Ohio. It is used by Chrysler Corporation to optimize the design of the outer body panels of Chrysler cars and trucks. SPAR's advantages are that it is interactive, easy to use, and fast.

SPAR and other useful software can be obtained at a moderate cost from NASA's Computer Software Management and Information Center (COSMIC), University of Georgia, Athens, Georgia. COSMIC maintains a library of over 1,300 computer programs applicable to a broad spectrum of business and industry operations.

A second example of a spin-off in the field of transportation is the Ride

Quality Meter, a diagnostic tool (meter), about the size of a breadbox that contains a computer, software, conditioning elements, sensors and meters for measuring vibration and noise levels, a liquid crystal display, and a printer. It measures and prints the ride quality of the vehicle being developed or evaluated when mounted on the moving vehicle.

The Ride Quality Meter was introduced to the commercial market in 1985 by Wyle Laboratories, Hampton, Virginia. It measures the discomfort level of a passenger subject to complex vibrations and noise. It is produced under NASA license and is based on a prototype meter and computer model developed by NASA's Langley Research Center, Hampton, Virginia. The Ride Quality Meter is being used by major manufacturers of transportation vehicles and parts (cars, buses, trucks, trains, aircraft, and spacecraft), including Ford Motor Company and International Harvester.

Energy

COSMIC produces a computer software package called PRESTO (Performance of Regenerative Superheated Steam Turbine Cycles), which is flexible enough to give a realistic prediction of design efficiencies. PRESTO is now part of many industrial energy systems, such as those of the Energy Systems Division of Thermo Electron Corporation, Waltham, Massachusetts, which specializes in the custom design of cogeneration systems (components of cogeneration systems include boilers, turbines, valves, generators, piping, and pressure regulators).

Another example is NASA's Remote Manipulator System (RMS) or Canadarm. Canadarm was developed in Canada as that country's contribution to the U.S. space shuttle program. It is a space shuttle-based crane that performs a variety of functions, operating as a 50-foot extension of an astronaut's arm either automatically or under manual control. The RMS weighs around 1,000 pounds and is capable of lifting 65,000 pounds (the equivalent of a fully loaded bus) in the weightlessness of space. Redesigned versions of RMS are now finding Earth-use utility in energy and mining uses, such as in remotely controlled underground mining equipment for improved safety and productivity.

New solid-state motor starters also evolved from space technology. They are designed to reduce energy consumption, lower maintenance costs, and extend motor life. The technology was originally developed at NASA's Marshall Space Flight Center (MSFC) in Alabama as part of NASA's energy conservation research in support of the Department of Energy. The Intellinet Corporation, Baltimore, Maryland, is one U.S. company that manufactures solid-state motor starters.

Consumer, Home, and Recreation

An attached (integral) part of the Apollo lunar suit worn by a dozen moonwalking astronauts was the suit's boots for cushioning and ventilation. The "spacer" material in the boots has turned up, in modified form, as a key element in a new family of athletic shoes designed for improved shock absorption, energy return, and reduced foot fatigue. Some of these athletic shoes are manufactured by Kangaroos, USA, St. Louis, Missouri.

The Fisher Pen Company, Forest Park, Illinois, makes several models of antigravity space pens now popular on Earth. The company developed the space pen for NASA astronaut record keeping on Apollo missions. (The USSR also purchased space pens from the company for use by its cosmonauts.) The pen was created to allow writing in orbit where ordinary pens, which rely on gravity and atmospheric pressure for ink flows, are inadequate.

Working under a NASA contract in cooperation with the Florida Solar Energy Center (FSEC), Dink Company, Alachua, Florida, has developed a prototype heat pipe dehumidification system that can double the moisture removal capacity of any air conditioner and save substantial amounts of energy. The project, when fully commercialized, should be a major step in controlling humidity in buildings, especially in hot, humid climates.

NASA also developed noise-abatement acoustic materials for use on its spacecraft. Variations of these acoustic materials have found their way into the marketplace as noise-deadening panels, sheets, adhesives, and enclosures used in aircraft, automobiles, homes, and offices. One U.S. firm that makes noise-abatement materials is SMART Products Company, Framingham, Massachusetts. The materials are known as SMART (sound modification and regulated temperature) products and are often 75 percent lighter than traditional soundproofing materials.

Industrial Technology

A high-temperature (duct) tape originally developed for NASA's use by the 3M Company is now being produced by 3M's Industrial Tape Division, St. Paul, Minnesota. The tape can withstand prolonged exposure to temperatures up to 500°F and is capable of functioning in temperatures as low as −65°F. Known as Scotch Brand Tape 364, it is an aluminized (glistening grey) glass cloth tape used in aerospace applications to protect electrical and instrumentation cables and fluid lines from rocket launch blast conditions. For nonaerospace applications, 3M Industrial Tape Division sees uses in the automotive and general transportation industries and for heat reflection applications in high-temperature building construction.

Research studies done by NASA and also on its behalf have been instrumental in the development of anticorrosion coating, which provides protection to steel ducts many times longer than epoxy coatings or polyester/flaked glass coatings. Anticorrosion coatings are used on steel pipes and ducts in industrial plants, homes, and offices.

Raymond Engineering, Middletown, Connecticut, designs, develops, and manufactures (among other products) specialized bolting and torquing equipment and tools for military, petrochemical, nuclear power, automotive, and other commercial applications. The company credits a NASA industry assistance center—New England Research Applications Center (NERAC), Storrs, Connecticut—with a supporting role in development of the bolting tools and in expanding the company's general technology base. Based at the University of Connecticut, NERAC is one of many NASA user assistance centers (Table 5.2) affiliated with universities across the country that provide information retrieval services and technical help to industrial and governmental clients.

Dodge Products, Houston, Texas, markets a line of solar energy sensing, measuring, and recording devices that incorporate solar cell technology first refined for space applications by NASA. Commercial customers include architects, engineers, and others engaged in construction and operation of solar energy facilities and manufacturers of solar-related products, such as glare-reducing windows (Taylor 1987, pp. 67–68).

Hohman Plating and Manufacturing, Dayton, Ohio, markets Surf-Kote C–800, a self-lubricating metal-glass-fluoride coating that resists oxidation at temperatures up to 1,600°F (870°C). Initially developed for missile and space uses, it is now used to protect sliding contact bearings, shaft seals for turbopumps, piston rings for high-performance compressors, and hot-glass-processing machinery (Taylor 1987, p. 67).

The list of spin-offs is enormous: rescue search beacons, long-lasting nickel cadmium batteries, electronic weight control systems for highway trucking, new metallic alloys, antifogging compounds for clearer windshields, freeze-dried foods, and hundreds of other devices and systems being developed, manufactured, and marketed throughout the world to improve and make more convenient our way of life. These spin-offs have given birth to a vast array of companies employing thousands of people.

There will continue to be many commercial spin-offs from space technology. Some will be beyond our wildest imagination and in new fields of endeavor. Companies interested in pursuing the licensing, commercial production, and marketing of NASA spin-offs should read NASA's *Tech Briefs* regularly and contact the various NASA field offices.

Tech Briefs, a monthly publication, contains information on NASA-related products and processes that is useful to industry, government, and academia. The publication may be obtained by contacting:

The Director
Technology Utilization Division
NASA Scientific and Technical Information Facility
P.O. Box 8757
Baltimore/Washington International Airport
Maryland 21240

A related publication deals with NASA-patented inventions available for licensing, which number about 4,000. NASA grants exclusive as well as nonexclusive licenses. A summary of all available inventions, updated semi-annually, is contained in the *NASA Patent Abstract Bibliography*, which can be purchased from the National Technical Information Service, Springfield, Virginia 22161.

NASA LOCATIONS

Tables 5.1, 5.2, and 5.3 list the names, addresses, and telephone numbers of NASA field centers, industrial application centers, and other important offices, respectively. For specific information concerning activities described in this chapter, contact the director of the specific NASA location or the appropriate technology utilization personnel at the address listed in the tables.

Field center technology utilization officers (Table 5.1) manage center participation in regional technology utilization activities. Industrial applications centers (Table 5.2) provide information retrieval services and assistance in applying technical information relevant to user needs. They are affiliated with universities across the United States. The Computer Software Management and Information Center (COSMIC) (Table 5.3) offers government-developed computer programs adaptable to secondary use. The application team (Table 5.3) works with public agencies and private institutions in applying aerospace technology to solution of public sector problems.

Information and assistance on space commercialization can be obtained from the Office of Space Commercialization at NASA Headquarters, Washington, D.C. (Table 5.3) and any other relevant NASA location. For information of a general nature about the Technology Utilization Program, address inquiries to the director of the Technology Utilization Division, Centralized Technical Services Group (Table 5.3).

SUMMARY

There are hundreds of spin-offs of space technology that have spawned new companies, created new markets, and benefited humanity socially and economically. This chapter presents a sampling of these spin-offs in various fields: public safety, health and medicine, transportation, consumer/home/recreation, and others. Some of these products are flame-

Table 5.1
NASA Field Centers

Ames Research Center
National Aeronautics and Space Administration
Moffett Field, California 94035
Technology Utilization Officer
Phone: (415) 694-5761

Goddard Space Flight Center
National Aeronautics and Space Administration
Greenbelt, Maryland 20771
Technology Utilization Officer
Phone: (301) 286-6242

Lyndon B. Johnson Space Center
National Aeronautics and Space Administration
Houston, Texas 77058
Technology Utilization Officer
Phone: (713) 483-3809

John F. Kennedy Space Center
National Aeronautics and Space Administration
Kennedy Space Center, Florida 32899
Technology Utilization Officer
Phone: (305) 867-3017

Langley Research Center
National Aeronautics and Space Administration
Hampton, Virginia 23665
Technology Utilization and Applications Officer
Phone: (804) 865-3281

Lewis Research Center
National Aeronautics and Space Administration
21000 Brookpark Road
Cleveland, Ohio 44135
Technology Utilization Officer
Phone: (216) 433-5667

George C. Marshall Space Flight Center
National Aeronautics and Space Administration
Marshall Space Flight Center, Alabama 35812
Director, Technology Utilization Office
Phone: (205) 544-2223

Jet Propulsion Laboratory
4800 Oak Grove Drive
Pasadena, California 91109
Technology Utilization Officer
Phone: (818) 354-2240

NASA Resident Office-JPL
4800 Oak Grove Drive
Pasadena, California 91109
Technology Utilization Officer
Phone: (213) 354-4849

National Space Technology Laboratories
National Aeronautics and Space Administration
NSTL, Mississippi 39529
Technology Utilization Officer
Phone: (601) 688-1929

Table 5.2
NASA Industrial Applications Centers

Aerospace Research Applications Center
611 N. Capitol Avenue
Indianapolis, Indiana 46204
Phone: (317) 262-5003

Kerr Industrial Applications Center
Southeastern Oklahoma State University
Durant, Oklahoma 74701
Phone: (405) 924-6822

NASA Industrial Applications Center
823 William Pitt Union
Pittsburgh, Pennsylvania 15260
Phone: (412) 624-5211

NASA Industrial Applications Center
Research Annex, Room 200
University of Southern California
3716 South Hope Street
Los Angeles, California 90007
Phone: (213) 743-8988

New England Research Applications Center
Mansfield Professional Park
Storrs, Connecticut 06268
Phone: (203) 429-3000

North Carolina Science and Technology Research Center
Post Office Box 12235
Research Triangle Park,
North Carolina 27709
Phone: (919) 549-0671

Technology Applications Center
University of New Mexico
Albuquerque, New Mexico 87131
Phone: (505) 277-3622

Southern Technology Applications Center
307 Weil Hall
University of Florida
Gainesville, Florida 32611
Phone: (904) 392-6760

NASA/UK Technology Applications Program
109 Kinkead Hall
University of Kentucky
Lexington, Kentucky 40506
Phone: (606) 257-6322

Table 5.3
Other NASA Offices

COMPUTER SOFTWARE MANAGEMENT AND INFORMATION CENTER (COSMIC)

Computer Services Annex
University of Georgia
Athens, Georgia 30602
Phone: (404) 542-3265

APPLICATION TEAM

Research Triangle Institute
Post Office Box 12194
Research Triangle Park,
North Carolina 27709
Phone: (919) 541-6980

COMMERCIAL SPACE PROGRAMS

Headquarters, National Aeronautics and Space Administration
Office of Commercial Programs
Commercial Programs Division
Washington, D.C. 20546
Phone: (202) 453-8430

SCIENTIFIC AND TECHNICAL INFORMATION FACILITY

Centralized Technical Services Group
NASA Scientific and Technical Information Facility
P.O.Box 8757
BWI Airport, Maryland 21240
Phone: (301) 859-5300

resistant fabrics, better breathing apparatus worn by firefighters, automated and faster systems for detecting drugs and chemical compounds in blood and urine, digital imaging, scalp coolers, advanced computer software, solid-state electronic devices, anticorrosion coating, and bolting and torquing tools.

The NASA locations listed in the chapter are valuable sources for keeping abreast of scientific and industrial inventions out of NASA, patent information, information about licensing NASA-developed or NASA-related products and processes, and new products and services that NASA is seeking. Many times the spin-offs find lucrative markets not only in the United States but in other countries as well. A company does not have to engage in space business directly to benefit from space ventures. However, firms that invest in space R&D are more likely than others to generate their own spin-offs. (R&D on spin-offs is generally proprietary information at large, well-known companies like Lockheed, IBM, 3M, and McDonnell Douglas.)

CHAPTER 6

SDI Is Big Business

In March 1983, President Reagan launched a new era in defense. He proposed the construction of a high-tech antimissile defense system that he said could shield the entire U.S. land mass and population and end the threat of nuclear war. President Reagan's proposed system, the Strategic Defense Initiative (SDI), more commonly known as Star Wars, is envisioned as a largely space-based defense system, with required ground-based support. SDI would employ kinetic-energy weapons (which destroy targets by physical impact), lasers, particle beams, and perhaps chemical weapons. The space-based weapons would include over 11,000 rockets on more than 2,200 orbiting battle stations or "garages" (Clausen and Brower 1987, p. 64) and other types of weapons, such as lasers. SDI is projected to cost over $2 trillion by the time it is fully deployed around the year 2000 or later (*Higher Education Advocate,* 1987, p. 1). It is big business for many U.S. corporations and overseas companies (*Houston Chronicle* 1985a, 1985c).

OVERVIEW OF SDI

At this writing, SDI is nominally still a vast research program with participants from federal agencies (such as the U.S. Department of Defense), large corporations (such as General Dynamics, Rockwell International, and General Electric), small companies with specialized skills (such as Geltećh, Inc., Alachua, Florida, and the Energy/Environmental Research Group, Melbourne, Florida) (Finegan 1987) and prominent universities (such as Stanford and MIT). SDI's most ardent supporters, including Caspar Weinberger, former U.S. secretary of defense, favor deploying an SDI defense system as early as the 1990s. SDI's critics, on the other hand, generally support the research, but they would pursue arms control with the Soviet Union and postpone the question of SDI deployment until more is known about the technologies and costs.

The policital motive behind early deployment is the desire to give SDI an irreversible momentum so that it cannot be tampered with by a future U.S. administration. Advocates fear that without a decision to deploy a space-based defense system—President John F. Kennedy's pledge to land a man on the Moon is invoked as a model—SDI will wither away once its patrons leave office. In the coming years, the United States will make critical decisions on SDI's budget, research strategy, and testing plans. Important questions will be debated. Does the United States need missle-defense research? Is SDI the right program to meet that need?

Some of the major evolutionary components of SDI are directed-energy weapons. They include orbiting chemical lasers, ground-based lasers, the free-electron laser (FEL), space-based kinetic-energy rockets (now called SBKKVs, an acronym for space-based kinetic-kill vehicles), the space-based neutral particle beam, and the U.S. Department of Energy's X-ray laser (Clausen and Brower 1987). Companies with these capabilities will find the following information useful for developing marketing and production plans for SDI projects.

Chemical Lasers

The SDI Organization (SDIO) oversees the SDI program. The budget for SDI increased from about $1 billion in 1984 to $3.5 billion in 1987 (Clausen and Brower 1987, p. 63). The SDIO initially focused on orbiting chemical lasers, which would derive power from heat-producing chemical reactions. These are technologically more mature than other lasers, but researchers have encountered many problems, including these three:

1. Tons of chemical fuel would have to be placed in orbit, at great cost, to power each laser.
2. Researchers conclude that no individual chemical laser could produce enough power to be an effective SDI weapon. Several lasers would have to be combined in phase, with their wave fronts marching in step to produce the 20 or more megawatts (million watts) of power required. The technology to accomplish this is in its infancy. While SDI continues to test a 2-megawatt hydrogen-fluoride laser-prototype called *Alpha,* few experts view the chemical laser as a promising space defense weapon.
3. Accurately aiming the comparatively long-wave infrared light from chemical lasers requires large, expensive mirrors.

Ground-based Lasers

The SDIO next turned its attention to ground-based lasers. These lasers, it is envisioned, will send beams to relay mirrors stationed 36,000 kilometers (22,360 miles) above fixed points on Earth, and the beams would be reflected from there down to "fighting" mirrors at about 1,000 kilometers (621 miles) of altitude. The most promising ground-based laser appears to be the FEL, first tested in the late 1970s. The electron beam that is caused to "wiggle" in a magnetic field produces the light in an FEL. Such a laser is potentially efficient and powerful, and its wavelength can be adjusted over a wide range to improve transmission through the atmosphere. The SDIO projects about $1 billion funding in 1987 and 1988 for FEL research.

The SDIO's goal for FEL research is to build a prototype weapon at the White Sands Missile Range, New Mexico, by the mid-1990s. The prototype would be a 100-megawatt laser that produces light at near-infrared wavelengths of about 1 micrometer (a millionth of a meter). This would be only about one-tenth the power needed for a practical weapon, but it is still far beyond present capabilities.

But before FEL can be an effective SDI weapon, some fundamental problems need to be solved. One is that the beam of the ground-based FEL would have to be pointed with much greater accuracy than has ever been achieved. The low-altitude fighting mirrors must be able to hold a beam 1 meter in diameter on a Soviet booster over 1,000 kilometers (621 miles) away that is traveling at up to 7 kilometers (4.3 miles) a second. When one booster is destroyed, the mirror must shift to another booster in a tenth of a second. An additional problem is that ground-based laser beams can be distorted by atmospheric turbulence and thermal blooming (changes in atmospheric properties caused by the heat the laser itself deposits in the air).

X-ray Laser

Another type of laser, now in its experimental research stage, is the X-ray laser. It would work by channeling the radiant energy from a nuclear explosion into thin rods. They would be vaporized to form a hot plasma whose ionized molecules would produce X-ray radiation. The radiation would be amplified along the axis of each rod, forming beams of potentially enormous power that could be aimed at several targets at once. The X-ray laser is envisioned as ground-based (where it is less vulnerable than being space-based) and would, in theory, be popped up upon warning of a Soviet attack.

But experimenting with nuclear explosions is an expensive process. Each nuclear test costs about $40 million, and the whole project consumes approximately $300 million a year. Although some tests have reportedly achieved lasing, the efficiency has been very low. The X-ray laser as a weapon will probably take at least fifteen years to develop, somewhere around the first decade of the twenty-first century. One of its most likely uses would be as an antisatellite weapon, since one device could attack several satellites at once. Ironically an X-ray laser in this role might well be turned against space-based military and communications satellites, with devastating effects.

Neutral Particle Beams

The SDIO is conducting research on space-based neutral particle beams. These beams would be generated by accelerating negatively charged ions to a very high energy, and then stripping the extra electrons off to form neutral atoms (probably hydrogen). Neutral beams do not quickly diverge the charged-particle beams, whose particles repel each other, nor are they bent by the Earth's magnetic field. Neutral beams of sufficient energy—for example, hydrogen atoms traveling at about 1 percent of the speed of light (at about 1,860 miles per second)—penetrate several centimeters into metals, making it impractical to shield against them (Clausen and Brower 1987, p. 70).

The particle beam is perhaps SDI's main hope for discriminating between warheads and decoys. An object illuminated by a particle beam emits radiation (gamma rays and neutrons) in proportion to its mass. By measuring radiation emitted by a warhead, a sensor could distinguish it from much lighter decoys. The SDIO plans to test this method on the space shuttle in the early to mid-1990s. The total cost of the experiments is estimated to be $1 billion.

Although the ideal appears feasible in principle, several hurdles need to be overcome before particle beams can be considered promising or reliable in practice. First, space-based electric generators to supply the space-based particle beam with the power it would need—about 100 megawatts, or enough to supply 50,000 homes—have not been developed. Second, the weight of current devices must be reduced by a factor of ten before they could be put in space at a reasonable cost. Third, they must be made far more reliable (e.g., with reliable, internal electronic devices) to operate unattended after years of storage in space. Finally, and perhaps most important, space-based beams are vulnerable in several ways. The Soviets could attack them and render them useless, use other particle beams to swamp the detectors with noise, or detonate nuclear explosions in space to blind the detectors for several seconds.

Space-based Kinetic-Kill Vehicles

The concept of space-based kinetic-kill vehicles (SBKKVs) gained currency in the early 1960s when the Department of Defense briefly funded, but then dropped, a similar project called *BAMBI* (ballistic missile boost intercept). In 1982, not long before President Reagan announced SDI, High Frontier, a space-defense advocacy group, championed the idea again. The Marshall Institute, a pro-SDI study group formed in 1984, urges early deployment of SBKKVs.

The SBKKV is a two-stage rocket weighing a few hundred pounds and carrying a vehicle with an infrared sensor to home in on a missile. Five to ten rockets would be carried on each space-based garage, orbiting at between 400 and 1,000 kilometers (248 to 621 miles) altitude to bring the rockets in range of Soviet missiles. The initial deployment outlined by the SDIO includes several hundred orbiting garages. A system to stop a large Soviet attack would perhaps require several thousand. Upon warning of an attack, the rockets would fire and accelerate to about 5 kilometers (around 3 miles) per second—nearly twenty times as fast as ordinary rocket interceptors like the *Exocet* antiship missile. Then they would home in on the bright flame of an intercontinental ballistic missile (ICBM) and destroy the missile by hitting it.

In 1987, Rockwell International was awarded a $75 million contract to design and test a prototype SBKKV (Clausen and Brower 1987, pp. 64–65). In September 1986, the SDIO conducted an elaborate $150 million test of technologies to track and intercept boosters, called the Delta 180 experiment (Clausen and Brower 1987, p. 65). Delta 180 showed that an interceptor could hit the third stage of a rocket by using radar, not infrared guidance. This is relatively easy since a radar beam bounces directly off the booster—not the flame—but radars are thought to be too heavy to be placed on a space-based rocket, and they can be jammed or fooled.

Ground-based Rockets

In addition to SBKKVs, SDI includes ground-based rockets to intercept Soviet warheads after they are released. One type would target them in the late midcourse, just before they enter the atmosphere, and the other would seek to stop them in the subsequent terminal phase as they drop toward targets on Earth. Infrared sensors carried on satellites or launched on rockets would track missiles and warheads by detecting their heat, and radars on the ground would be used as well. No new physical principles need to be understood to make these kinds of weapons work, but many scientists think the prospects of these weapons offering an effective missile shield are dim. Some of these scientists offer several arguments against SDI.

ARGUMENTS AGAINST SDI

Critics have advanced many arguments against SDI, from technological problems to budgetary issues. Six are important:

1. Space-based rockets—and space-based defenses of all kinds—are very vulnerable to attack by Soviet antimissile (ASAT) weapons and similar weapons of other enemy nations. A particularly simple antisatellite weapon, for example, is a space mine, which would dog the heels of a defensive satellite until ordered to detonate and destroy the rocket(s) or garages in which they are housed.
2. Many engineering problems in SDI are yet to be solved; some critics feel that SDI is technically infeasible. Most of these problems focus on the difficulty of developing rockets with pinpoint accuracy for hitting fast-moving, attacking missiles or being overwhelmed by decoys and/or real warheads.
3. Since the rockets would orbit at about 1,000 kilometers (621 miles) of altitude, only a fraction would be in position to fire on Soviet ICBMs at any given time. This so-called absentee problem means that a large number of U.S. rockets must be deployed for each missile to be intercepted. Clausen and Brower (1987, p. 67) believe that the ratio will be about fifteen to one for the Soviets' missile force of the mid–1990s if they continue to base their missiles at fields scattered across the country. If the Soviets tightly cluster their missiles, the absentee problem may be worse; as many as 100 rockets may be required for each ICBM.
4. SDI will serve to inspire or stimulate Soviet nuclear arms buildup, with potentially disastrous consequences.
5. There are concerns about the reliability and accuracy of the computer software required to run SDI. The SDI computer program will require at least 10 million lines of instruction (*Higher Education Advocate* 1987, p. 1) and numerous other software for ground-based and space-based command posts and for key guidance systems aboard missiles.
6. The SDI project will drain the United States of funds that could be better spent on scientific and medical research, and on economic development.

In sum, major arguments against SDI are technical infeasibility, vulnerability to attack, the absentee problem, Soviet countermeasures, computer software problems, and budgetary difficulties. But there are also strong arguments in favor of SDI research and eventual deployment.

ARGUMENTS IN FAVOR OF SDI

The most potent argument in favor of SDI is that the Soviets are developing SDI capabilities and countermeasures; the United States should do the same for its own safety and as a strong deterrent against a first strike by the Soviets. Other supporting arguments in favor of SDI are as follows:

1. SDI will contribute to national security.
2. SDI perfection is not needed to provide an effective counter to offensive forces. The ability to destroy 90 percent of attacking missiles would probably deter an enemy. Current U.S. offensive systems are imperfect, but they have provided protection for several decades.

3. Although substantive engineering issues remain unresolved, including the basing choices for the weapons (e.g., different orbits, and different land bases) and design options that would enable enough weapons to survive an attack, these engineering problems are not insurmountable.
4. SDI will not rely on a few types of weapons but several, and these will be in abundance to strengthen attack and counterattack capabilities. For example, there will be ground-based and space-based kinetic-kill weapons, as well as orbiting laser weapons that could attack at the speed of light and therefore operate at higher altitudes than kinetic weapons. Morrow (1987, p. 25) believes that this would enhance SDI capabilities and survival from attack.
5. The budgetary cost of SDI should not be an issue given the importance of national defense and security.

In sum, the major arguments in favor of SDI are that the Soviets are also engaging in SDI research, SDI is important to national security, and SDI will deter a Soviet first strike. Other arguments that have been articulated for SDI research are increased employment, spin-offs into the domestic marketplace, and general technological advancement that will result from SDI contracts awarded to corporations and universities.

PROFITS IN SDI

SDI research has created bonanzas for many American companies and universities and will continue to do so for many years to come. Regardless of the wisdom or feasibility of SDI, these organizations see SDI's anticipated $19 billion, five-year research budget (Finegan 1987, p. 68) as a once-in-a-generation opportunity. Whether it is deployed or not, Star Wars promises to push back the frontiers of science and technology in ways reminiscent of the race to the Moon. Companies that get in on the ground floor can look forward not only to years of lucrative government contracts but also to the potential of huge profits from commercialization of technology developed at government expense. Profits from SDI projects and spin-offs will amount to millions of dollars per year for successful participants.

Company Beneficiaries

Among the companies that have already been awarded SDI contracts or will get such contracts because of their tremendous track record and expertise are General Dynamics, Martin Marietta, Rockwell International, Lockheed, McDonnell Douglas, TRW, Hughes Aircraft, General Electric, and IBM. By and large, the value and stocks of these companies have been enhanced over the years by lucrative defense contracts (see Govoni 1986b, p. 26, for example). Other likely candidates for SDI contracts include Grumman Aerospace, Honeywell, RCA, Westinghouse Electric, Frequency Electronics, General Defense, and Alpha Industries. Smaller R&D firms such as Helionetics (solar energy, laser technology, electronics), Optelecom (fiber optics and laser systems), Irvine Sensors (infrared-detection systems), and Helix Technology (cryogenic-refrigeration equipment) have also been mentioned (Govoni 1986a).

Of the hundreds of small companies participating in SDI research, some are teamed with big corporate contractors such as General Dynamics or Rockwell International; others have earned SDI contracts through specialized expertise and nichemanship. Included in this latter group are Geltech, Inc., Alachua, Florida; Energy/Environmental Research Group (EERG), Melbourne, Florida; Sparta, Huntsville, Alabama; Nichols Research Corporation, Huntsville, Alabama; W. J. Schafer Associates, Arlington, Virginia; and Systems Planning Corporation, Roslyn, Virginia (Finegan 1987).

Geltech, built around a new technology for making glass, is typical of the small firms with special expertise that won SDI contracts. Rather than melting sand in the traditional method, Geltech has found a way to mix silica-based solutions that harden into glass that the company claims is purer and stronger than ordinary glass and more easily fashioned into large pieces than anything previously attainable. SDI will need many large, lightweight mirrors to bounce directed energy beams at missiles. Geltech is playing a central role in that research with a $1.6 million sixteen-month contract (Finegan 1987, p. 70), with possibilities for more contracts in the future. Its new method for producing glass may permit missiles to be shot down by laser beams bounced off huge, lightweight mirrors in space.

EERG was launched in 1980 in Tucson, Arizona, by University of Arizona astronomy professors John Scott, Eric Craine, and a third colleague as a vehicle for consulting work. One of the problems facing SDI engineers is that the satellite sensors ("eyeball") that look down on, say, the Soviet Union have limited peripheral vision and produce imprecise images. It is possible that they might miss the beginning of a missile attack. In 1985, EERG patented a technique, under the name of Rimstar, that offers the possibility of a fixed infrared "eyeball" that could stare down at large geographic areas with near-perfect vision.

EERG's three-year contract, worth $4 million to $5 million, will take Rimstar into uncharted territory. Should EERG succeed in designing an infrared optical electronic sensor (camera) capable of withstanding the rigors of space travel and attack from Soviet missiles, the company will be in line for other, much larger contracts, such as for development of SDI's early warning and missile-tracking system. These are frontline components on which the entire SDI program will depend. There is also the possibility of many other military and civilian applications of the infrared camera being developed by EERG. They include helicopters, tanks, TV-guided missiles, and commercial and private aircraft for better navigation at night or in bad weather.

Sparta had 335 employees and sales of $32 million in 1986. The company is known for its sophisticated computer modeling and has won many computer-related defense projects in the past. As one of the principal designers-architects for the SDI system, Sparta must find a way to integrate thousands of new technologies in a fail-safe defense system. It must provide the blueprint by which others do the brick-and-mortar work of manufacturing and assembling the components of the system. Industry analysts believe the architecture contract for SDI could be worth $20 million to $50 million a year. Even after full SDI deployment, Sparta is likely to have plenty of work incorporating new technologies designed to blunt Soviet countermeasures.

In its SDI work, Sparta is teamed with Nichols Research Corporation, a specialist in optical analysis with 430 employees and $27 million in sales; and W. J. Schafer Associates, a $16 million company that has already won SDI contracts for its expertise with lasers. These partners understand the technical issues and have successful track records. In 1986, Sparta's SDI projects totaled $24 million.

University Beneficiaries

Research related to SDI brings in millions of dollars to such universities as Georgia Institute of Technology, Massachusetts Institute of Technology, Stanford University, and the University of Texas at Austin.

In 1985, Georgia Tech won a $21.3 million contract for work on high-speed computers to help solve problems with missile interceptions. The goal of the project, based at the School of Engineering, is to build a special-purpose computing system that could be used by the military for guidance and control (of missiles) or by civilians for complicated engineering problems (Corddry 1985).

Engineers and researchers at the University of Texas (UT) are working on a rail gun that could be used to shoot down Soviet nuclear missiles or hurl 1-ton cargoes into orbit around the Earth. Both the weapon and the cargo launcher would be powered by a unique high-energy "compact homopolar" electrical generator developed by UT engineers. The SDI program envisions putting rail guns as long as 25 feet into LEO. The rail guns would be capable of accurate sixty shots a minute at ranges of about 600 miles (*Houston Chronicle* 1985a).

As of mid-1987, over 3,000 Pentagon contracts, for more than $6.7 billion worth of space-defense research (Cunningham 1987), had been awarded to large companies, small companies, and universities. These companies and universities often collaborate on research. Beyond the obvious benefits of these contracts (profits, increased employment, and income), SDI will also spawn worldwide spin-offs that will create market opportunities and wealth for many organizations.

SDI Spin-Offs

Some of the commercial spin-offs from SDI will be beyond our wildest imagination; others are fairly obvious. The latter should include advancements in communications, transportation, energy generation and utilization, optics and image processing, new high-strength, lightweight materials and structures, data handling and processing by more sophisticated computers and software, and radar. American businesses must be poised to take advantage of these opportunities. Even if the SDI program ultimately falls short of its ambitious objectives, the research that will have been lavished on it is likely to generate sufficient spin-offs to fuel innovation and advances in other fields well into the twenty-first century. Organizations working directly on SDI projects, however, will encounter a few hurdles.

THE PENTAGON AS PARTNER

If a company has an excellent track record and sells products and/or services that the SDI program needs, getting an SDI contract may be the easy part. Managing it is another story. Dealing with the U.S. federal government invariably involves a great deal of red tape and subsequent delay. There is even more red tape and delay with SDI work, partly because of the need for security checks and clearances on companies, present employees, and the newly hired who will be doing SDI work. Other troublesome areas are allowable profit levels, penalties for fraud, and subcontracting.

Because a great deal of SDI work is classified, almost every company that obtains an SDI contract has to follow security procedures. This complicates business enormously. Not only must detailed security measures, described in thick manuals, be adhered to, but security clearances for new employees sometimes take six months or longer, and they cannot be submitted for clearances before they are hired. In the meantime, they cannot work on what they were hired to do—creating a payroll problem and a morale problem, as well.

With respect to profits, the Pentagon—reacting to budget pressures and criticism that defense contractors make too much money—issued new guidelines on allowable profits. A research operation, with little invested in capital but a lot invested in people may have an allowable profit level lower than that of an airplane manufacturer with a large investment in tooling. In fact, the allowable profit level could be as low as 10 percent before taxes. The federal government has also become tighter on reimbursement for travel expenses. Under new travel guidelines, for instance, contractors may claim no more than the per diem rate allowed government employees. But government employees often get significant discounts at major hotel chains not available to corporate customers. Consequently some corporate customers may end up subsidizing their travel expenses.

The Pentagon is now imposing more severe penalties (e.g., jail sentences and large fines) on corporate officers for fraud. Similarly, subcontracting is subject to the government's strict purchasing laws. For example, often if the subcontracting firm does not take the lowest bid price, it has to go through an elaborate justification, and government auditors may audit all the firm's bids relative to the SDI project(s).

These issues of security, profit levels, penalities, and subcontracting are some of the decision areas about which SDI contractors should be aware. SDI contracts can be very profitable, but SDI contractors face bureaucratic guidelines not common in the private sector. The more prepared contractors are for these guidelines, the smoother will be their working relationship with the federal agency involved.

SUMMARY

SDI is lucrative business for many well-known corporations like General Dynamics, Rockwell International, and IBM; for many smaller companies with specialized expertise; and for many universities. Over $8 billion worth of SDI contracts have already been awarded (up to 1988) with an estimated average of between $10 billion and $20 billion worth of SDI defense contracts annually for at least the next ten years. One estimate is that the U.S. government will spend about $2 trillion on SDI by the time it is deployed, around the year 2000.

These expenditures are for goods and services such as ground-based and space-based rockets, ground-based and space-based lasers, defense posts in space, launch vehicles and space transportation systems, surveillance satellites, orbiting storage facilities with different types of weapons, computer hardware and software, basic scientific research services, pilot tests of weapons for accuracy, potency, and so on, and counter-SDI research.

Some of the arguments advanced against SDI are technical infeasibility, vulnerability to attack, the absentee problem, dangerous arms built by the United States and USSR, and budgetary problems. Arguments in favor of SDI include the fact that the Soviets are engaging in SDI research and countermeasures and so, too, should the United States, the importance of SDI to national security, and SDI as a deterrent to a Soviet first strike.

Many corporations and universities will make large profits from SDI contracts. These contracts will generate tremendous employment opportunities for years to come, incomes, an economic multiplier of three or more, economic growth, technological advancements, such as in telecommunications, radar, and space medicine, and spin-offs in many fields. The profits to be earned from SDI are not without attendant difficulties, however. These difficulties include adherence to strict security procedures, FBI and other security checks of persons working on SDI projects, and restricted profit levels for organizations working on SDI.

Regardless of the occasional political overtures between the United States and the USSR and the veneer of harmony, the nations do not trust each other. SDI research is necessary for the well-being of the United States and as a deterrent to Soviet attack. It is here to stay. U.S. corporations and universities must be prepared to take advantage of the bonanzas that SDI is creating—in research, in the production and marketing of defense systems to the U.S. government, and in lucrative spin-offs that are bound to occur in the marketplace.

CHAPTER 7

A Methodology for Identifying Customers for In-Orbit Facilities

The major countries involved in space commercialization are the United States, the USSR, Japan, China, Italy, members of the ESA, and Canada. Within these countries, there are many well-known companies involved in space commerce. RCA (satellites), IBM (computer hardware and software for space R&D), General Dynamics (satellite launch systems, space weaponry), British Aerospace (launch vehicles), and Aeritalia, the Italian aerospace contractor that built the spacelab hardware for ESA. One of the biggest problems facing some of these companies is how to identify private sector customers (companies) for target marketing. Public sector customers (government space agencies) are easy to identify, they include NASA, NASDA, ESA, and the Ministry of Astronautics in the PRC.

The purpose of this chapter is to describe and illustrate a methodology that many companies (and governments too) in the space industry can use to help identify private sector customers for in-orbit facilities on space stations and space platforms. These facilities include laboratories, workspace, and utilities and are an important part of the space infrastructure, especially for materials processing in space. NASA, NASDA, ESA, Japan, and Canada are each starting the process of trying to identify customers for the U.S./International Space Station (Foley 1987a, p. 24).

Customer identification is central to effective, efficient, and profitable marketing and also important in setting production and inventory levels. Since space business is very expensive, a company cannot afford to waste time and money spinning its wheels in the marketplace, trying to sell its products or services to unidentified customers. In short, customer or market identification is the first key to successful marketing. Other keys include a quality product at a reasonable price, effective implementation of good marketing programs, analysis of competitor strategies, counterstrategies, and excellence in customer service.

OVERVIEW

The methodology described here focuses on one of the most difficult tasks in space commerce: identifying private sector customers for in-orbit services on space facilities (a term that encompasses space platforms and stations). These services include the provision of laboratory facilities for MPS, as well as for a variety of space science, physics, and Earth observation experiments; commercial workspace in space; and utilities, such as electricity.

In space, these services are provided to a limited extent by the U.S. space shuttles, space stations, and space platforms. The space shuttle is of limited use for MPS for at least three reasons: its space is limited; its flight or mission duration normally lasts seven to ten days, much too short for most MPS activities, such as biological and industrial production; and crew motion and interaction cause a less-than-ideal environment for microgravity-sensitive MPS payloads.

With respect to space stations, NASA, ESA, Canada, and Japan are collaborating in the development of the U.S./International Space Station. It is being designed as a permanent research, manufacturing-processing, satellite-repair, and spacecraft refueling facility in LEO around 250 to 300 miles above Earth. It will be assembled in space between 1995 and the year 2000 by way of twenty to twenty-five shuttle flights and will have various expandable modules and systems, such as power supply, laboratories for MPS and other activities, living quarters for about six to eight astronauts and scientists, and on-board computers for data analysis and management.

Permanently staffed, human-tended and automated space platforms are also being developed by various firms and countries. Permanently staffed space platforms will have a crew on board (perhaps two to four), life support, logistics supply, and environmental control systems. Human-tended space platforms are being designed with docking or berthing modules for docking with the U.S. space shuttle or U.S./International Space Station for life support systems. Automated space platforms generally have no compartments for human habitation and are often called free flyers. Examples of space platforms are ESA's European Retrievable Carrier (EURECA), the SPAS, built in West Germany, and the ISF being developed by Space Industries, Houston, Texas. General Electric, RCA, and Fairchild Industries are also building space platforms. Generally space platforms are rectangular, square, or cylindrical, with length, width, height, or diameter about 15 to

25 feet. They are laden with laboratories and instrumentation and can cost $300 million each and up, depending on factors such as size and on-board facilities.

Permanently staffed facilities, such as a space platform or a space station, will be required for operations where constant interaction with the payload is required. Examples are basic research into biological phenomena, development of new process technologies, and performance of processes that require complex handling (such as repacking and refrigeration of biologicals). Man-tended facilities (MTFs) provide a "shirtsleeve" working environment for relatively short periods of time. They would be staffed while the production processes ran. They could be docked with space shuttles or space stations. The advantages of the MFT over a permanently staffed facility are lower costs and better microgravity conditions (experiments can run for longer duration with less disturbance.)

Automated free-flying platforms are most economical production sites for processes that have reached a high level of maturity or are extremely simple. These turnkey systems could be launched on a free-flying platform and left untended for months, with production batches picked up by a space shuttle or other spacecraft. Automated free-flying platforms may be cheaper to build and operate than human-tended or permanently staffed facilities.

There is not a great deal of demand for in-orbit services for these reasons:

1. Financially feasible and proved MPS products or processes that would use such services do not currently exist.
2. The products that have been identified with possible production in space, (e.g., some pharmaceuticals, some metal alloys) are not financially attractive investments now.
3. Transportation to and from space is extremely costly. One estimate is $10,000 per kilogram (*Economist* 1984b).
4. Increasing attention is being paid to Earthbound production techniques that might do just as well as space-based production techniques.
5. Governments control access to space, and this has restricting implications for private industry.
6. Markets for space products are underdeveloped.
7. Private businesses as a whole are uneducated or "space naive" about the potential of the space environment.
8. The financial markets are reluctant to lend money to companies to invest in what these financial institutions perceive as very risky ventures.
9. There are tremendous difficulties in obtaining space insurance, and at a reasonable cost.

In spite of these obstacles, there will be a demand for in-orbit services as the U.S./International Space Station, other space stations, and space platforms come on line, and as MPS and other space activities become more commonplace.

The following are unique characteristics of space platforms in general. They must be kept in mind because they have an important bearing on the methodology of customer identification for leasing commercial workspace and utilities on these space facilities:

1. Space platforms are new products, so manufacturers have relatively little or no experience regarding how to identify customers.
2. They are very expensive assets, generally costing $300 million or more, depending on size, type, complexity, and amount of technological equipment. These are construction costs only, not operating and maintenance costs.
3. They are being designed to have a useful life of ten to fifteen years. (Straight line depreciation could be used for writing off the asset.)
4. They may be available on a service contract basis or may be sold outright to some customers.
5. They may be capable of producing up to 20 kilowatts of sustainable electrical power.
6. If we assume the following—a manufacturing cost of $300 million, a launch cost of $100 million, a ten-year lifetime, and straight line depreciation—the annual fixed cost is roughly $40 million.
7. The MTF or space platform requires servicing every three months by a space shuttle. The U.S. shuttle stays in space an average of ten days, and MTF servicing may take an average of three days per MTF. Accordingly, the shuttle or similar space transportation system (STS) could service as many as three MTFs on a single flight. If the STS flight costs $100 million ($400 million for four flights per year), the yearly transportation-servicing cost of each of three MTFs is roughly $133 million. Total yearly "fixed" and "variable" costs in this case are $173 million. If the MTF has 20 kilowatts of power, this yields a cost of almost $9 million per kilowatt year.

THE CUSTOMER IDENTIFICATION PROCESS

The steps for identifying customers are divided into two board categories: internal and external. Many of the steps are carried out concurrently, and some are not clearly internal or external but imply attention to both aspects.

Internal Steps

Understand Customer Needs. In order for the owner of the space facility to succeed in leasing (or selling) utilities and commerical workspace, the company must thoroughly understand its potential customers and what they precisely need and want. This is a catch–22 situation. The builder-owner is trying to identify potential customers for a new and unique product or service (a difficult task at best) and customers often do not know what they want in the space environment (recall that many are space naive). This makes customer identification more arduous and challenging.

Three methods are suggested for understanding the space needs of potential users of the space facility, such as MPS firms:

1. Examine case stories or literature on other makers of space platforms in the United States and elsewhere to try to locate information about the needs and wants of potential customers. Sources of case stories include *Aviation Week & Space Technology, Space World,* the *Wall Street Journal,* the *Economist,* and *Science Digest.*
2. Organize visits to top executives in well-known space firms. Discuss customer needs and customer identification issues with them. (Given the proprietary nature of space research, organizing such visits and meetings may be difficult.)
3. Attend conferences on space commercialization that are held in the United States and other parts of the world. NASA sponsors some of these conferences in Houston, Washington, D.C., Los Angeles, and other major U.S. cities. Contact NASA for such information. Conference information, such as location, dates, and major topics, also appears regularly in *Aviation Week & Space Technology.*

These sources will help in understanding customer needs in terms of in-orbit utility requirements, laboratories, instrumentation, computers, and data management.

Jury of Executive Opinion. This technique involves soliciting the judgment of a group of experienced managers of the organization to identify potential customers for the space facility. These judgments are often solicited through a few meetings, using formalized written reports or informal brainstorming procedures. These might include forecasting techniques, such as the delphi method, which emphasizes emergent and future customer needs. The main advantages of the jury approach are that it is fast and allows the inclusion of many subjective factors, such as competition and economic climate. The continued popularity of this method of forecasting and identifying customers shows that most managers prefer their own judgment to other less well-known statistical forecasting procedures.

"Kitchen Cabinets." Kitchen cabinets are informal consultants, who often bring a fresh viewpoint to problems that stymie managers. Potential members are retired executives from similar businesses (e.g., the aerospace industry and NASA), professors from local universities (such as those in computer science, engineering, mathmetics, and business), subcontractors, customers, other professionals, and a few key internal corporate executives. Kitchen cabinets can brainstorm issues, bring their industrial and scientific experiences to bear on the problem(s) at hand, and engage in synergistic and creative thinking. At least five inquiring modes have been identified: synthesist, realist, idealist, pragmatist, and analyst (Harrison and Bramson 1982). Representatives of each style of thinking tend to ask different questions premised upon distinctive preferences and perspectives. Each approach can yield provocative and complementary insights.

The kitchen cabinet should meet quarterly and concentrate on one or two issues facing the company. Hidden benefits are generation of good ideas, an inexpensive method of consultation, and recommendations by members of a kitchen cabinet to potential customers.

Fortune 500 Companies. The list of Fortune 500 companies is perhaps the richest data base to start the search for potential customers interested in MPS and other space activities. Select from the list companies in the industries that can potentially make use of unique characteristics of the space environment: microgravity, vacuum, sterile conditions, lack of vibration, unlimited heating and cooling, lack of atmospheric attenuation, viewing capability/perspective, and radiation. For MPS activities, the microgravity attribute is of greatest importance. MPS industry applications include processing biological materials, high-performance computer crystals, ultrapure glasses, and metal alloys ten times stronger than steel but lighter.

After this initial pruning of the Fortune 500 list and classifying companies into major, relevant industries, a final elimination from the list should be made of companies that spend under $200 million a year on R&D. This benchmark is established based on the fact that MPS will require at least $200 million in research development funds.

Tables 7.1 and 7.2 list the major industries and companies, respectively, that are potential customers for MPS. The tables, though quite complete, are not meant to be exhaustive; they serve only to point out the manner in which a developer of a space platform or space station must locate customers.

Some of the companies listed in Table 7.2 may qualify as customers as well as competitors. Add to this list of commercial companies the U.S. Department of Defense, NASA, and foreign governments and corporations, and the developer of the space facility is looking at a challenging customer mix. The developer must decide early which is the selected market: the government, the private sector, or both. This choice has obvious implications for the space facility design, commercial and/or military uses, customer mix, and so on. The nature of the high-risk, high lease price for use of the space facility (variously estimated at $2 million to $4 million per month), a long payback period (eight years and beyond) for both the customers and the owner of the space facility, and the potential for a mix of customers, points clearly to a monumental marketing effort.

External Steps

Contacts with NASA. Assuming that the owner of the space facility is in the United States, that company has to develop a close working relationship with NASA to understand government guidelines for designing and building the space facility; obtain names, addresses, and telephone numbers of NASA's primary contractors (who may be potential customers of the space facility); and keep abreast of technological developments in space commerce. Contacts are established through past performance as a NASA contractor, personal visits with top NASA personnel, reading *Tech Briefs,* and so on.

Market and Financial Analyses. Doing some market and financial analyses for customers is almost imperative for identifying potential customers. Let us assume that these customers know how to identify some of their own customers (market), perhaps using the methodology described in this chapter. However, most potential customers for the space facility do not have the details of the financial costs of leasing utilities, laboratories, or workspace on the space facility. The manufacturer or owner of the facility must provide these costs.

These costs should be realistic, not optimistically low. They should include round trip transportation costs to the space facility, number of trips per year, cost of any equipment the customer carries to the space facility, leasing costs of the space facility, salaries of scientists or payload specialists sent up by the leasee, space insurance costs (for space hardware and the workers in space), salaries for ground-based support personnel, the cost of using ground-based assets for the space venture, and the cost of capital.

Knowledge of these costs is imperative for potential customers so they can evaluate the nature of returns on, say, any given MPS investment. Potential customers who do not see a substantial return on investment will not invest in MPS and will not lease or buy workspace on the space facility. In short, the owner of the space facility will make profits only if potential customers perceive that they too can make substantial profits on their space investments.

While it is not the intent of this chapter to spell out the details of a financial model that potential customers can use to help them determine whether to invest in the MPS activity, a few comments along these lines are instructive.

One can use a mainframe financial planning system, such as IFPS (Interactive Financial Planning System) for extensive financial modeling and analysis, or a more modest financial model using Lotus 1–2–3 software on an IBM PC-XT. Many other software packages are available for financial analysis.

The financial model is based on cash flow estimation methods used in capital budgeting theory. Typical input data are customer number, land, building, and equipment costs, the amount of R&D costs to be capitalized, depreciation (building, equipment), expected sales in units per year, sale price per unit, the cost of capital, estimated taxes, and so on. The model can calculate total investment required, cash flows, net salvage values, the time line of consolidated cash flows, payback period, net present value

Table 7.1
Potential Customer Industries for Space Facilities

Industry Code for this Chapter	Industry	SIC Number
01	Foreign Govts. & Corps.	NA
02	U.S. Department of Defense	NA
03	NASA	NA
1	Aerospace	3861,3764,3769
2	Chemicals (Industrial)	
	Chemical and Allied Products	2861,2865,2869
3	Computers and Office Equipment	3573
4	Drugs and Medical Products	2831-34
5	Electronics and Appliance	3671-79
6	Measuring, Scientific,	3811,3822
	and Photographic Equipment	3861
7	Medical Equipment	3841-43
8	Metals-Nonferrous	3361-69
9	Motor Vehicles and Parts	3714
10	Steel	3312
11	Telecommunications	4811

Table 7.2
Potential Customer Companies for Space Facilities

Company	Fortune 500 Rank	Industry Code (From Table 7.1)
Allied Corp. (Morris Township, N.J.)	26	2
American Can (Greenwich, Connecticut)	124	8
American Telephone & Telegraph (New York, N.Y.)	8	5,11
DuPont (E.I.)de Nemours (Wilmington, Del.)	7	2
Eastman Kodak (Rochester, N.Y.)	28	6
Eli Lily (Indianapolis, Indiana)	130	4
General Electric (Fairfield, N.Y.)	9	5,7
International Business Machines (Armonk, N.Y.)	6	3
Johnson and Johnson (New Brunswick, N.J.)	57	4,7
McDonnell Douglas (St. Louis, Missouri)	34	1
Minnesota Mining & Manufacturing (3M) (St. Paul, Minnesota)	45	3
Monsanto (St. Louis, Missouri)	51	2
Rockwell International (Pittsburgh, Pa.)	37	1
Smith Kline Beckman (Philadelphia, Pa.)	136	4
Texas Instruments (Dallas, Texas)	63	3
United Technologies (Hartford, Connecticut)	16	1
U.S. Steel (Pittsburgh, Pa.)	15	10

(NPV), and the internal rate of return (IRR). The output parameters (the payback period, NPV, and IRR) can then be used by the space facility's potential customers to make a judgment about the investment.

Assumptions used in the model will affect the outcomes. Some of these assumptions include the cost of capital, the rate of depreciation on building and equipment, the inflation rate, and the amounts of the R&D investment that will be partially expensed and partially capitalized. The amounts depend on particular tax laws, as well as required financial results a company must achieve.

International Trade Fairs. Trade fairs are intended to provide commercial exposure for the products of the exhibitors. They are generally held in large exhibition halls or pavilions in major cities. At these fairs, the space facility manufacturer should have engineering mockups of the space facility and its various components, or actual parts, audiovisual descriptions of the facility on large television screens, colorful brochures and booklets for distribution, and scientists and engineers on hand to answer questions about the facility. The brochures and booklets should contain the names, addresses, and telephone numbers of contact persons at the manufacturing company and should address typical issues and questions asked by potential users of outer space for MPS and scientific studies. Representatives of the space facility manufacturer must also make sure to collect the names, addresses, and telephone numbers of potential customers and to follow up shortly after the fair.

Examples of well-known international trade fairs that may be of interest to makers of space facilities are the Paris Air Show, the Farnborough Air Show (held in England), and others in Leipzig (electronics), Frankfurt (computers, robotics), and Dallas, Texas (microcomputers).

SUMMARY

Space business is big business—more than $100 billion annually. This money is spent on the production and marketing of space goods and services like communications, military and weather satellites, space shuttles, space stations and space platforms, defense systems, and transportation into space.

Many countries are involved in space commerce: the United States, the USSR, Japan, the thirteen-member countries of ESA, the PRC, Italy, and Brazil. Within these countries there are many firms involved in space ventures.

Companies that make space stations and space platforms (space facilities) may have difficulty identifying potential customers. This chapter discusses a methodology useful in identifying private sector customers who seek to lease or buy in-orbit utilities, laboratory facilities, and workspace on space facilities (emphasis on space platforms).

After a description of the space facility characteristics, the methodology is described. Internal steps in the customer identification process include understanding customer needs, use of a jury of executive opinion and kitchen cabinets, and, most important, a detailed analysis of Fortune 500 companies. An example of the last is presented. The three external steps in the customer identification process are to establish contacts with NASA, provide some financial cost analysis for potential customers, and attending international trade fairs.

Potential customers are frequently those already engaged in space ventures. They tend to be large, wealthy, internationally known corporations, such as McDonnell Douglas and 3M, and are often already engaged in MPS and other space experiments. They are from the aerospace, nuclear energy, biotechnology, engineering, electronics, and telecommunications industries. These potential customers will not become actual customers unless it is profitable for them to use in-orbit facilities for MPS or scientific and industrial studies and unless they have large enough markets on Earth for the fruits of their space efforts. So far this has not been the case. Other impediments to leasing utilities and workspace in orbit include expensive space transportation, difficulties in obtaining space insurance for expensive assets and for people working in space, long payback periods (usually eight years and beyond), and complex government regulations. Additionally, owners of space facilities for lease or purchase by the private sector face a monumental task of marketing their space facilities and in-orbit services to potential customers after identifying these potential customers.

Space business is long term and risky. However, entrepreneurs who are prepared to begin the process of lifting their sights beyond terrestrial markets may be among those best positioned to become the commercial superstars of the twenty-first century and beyond.

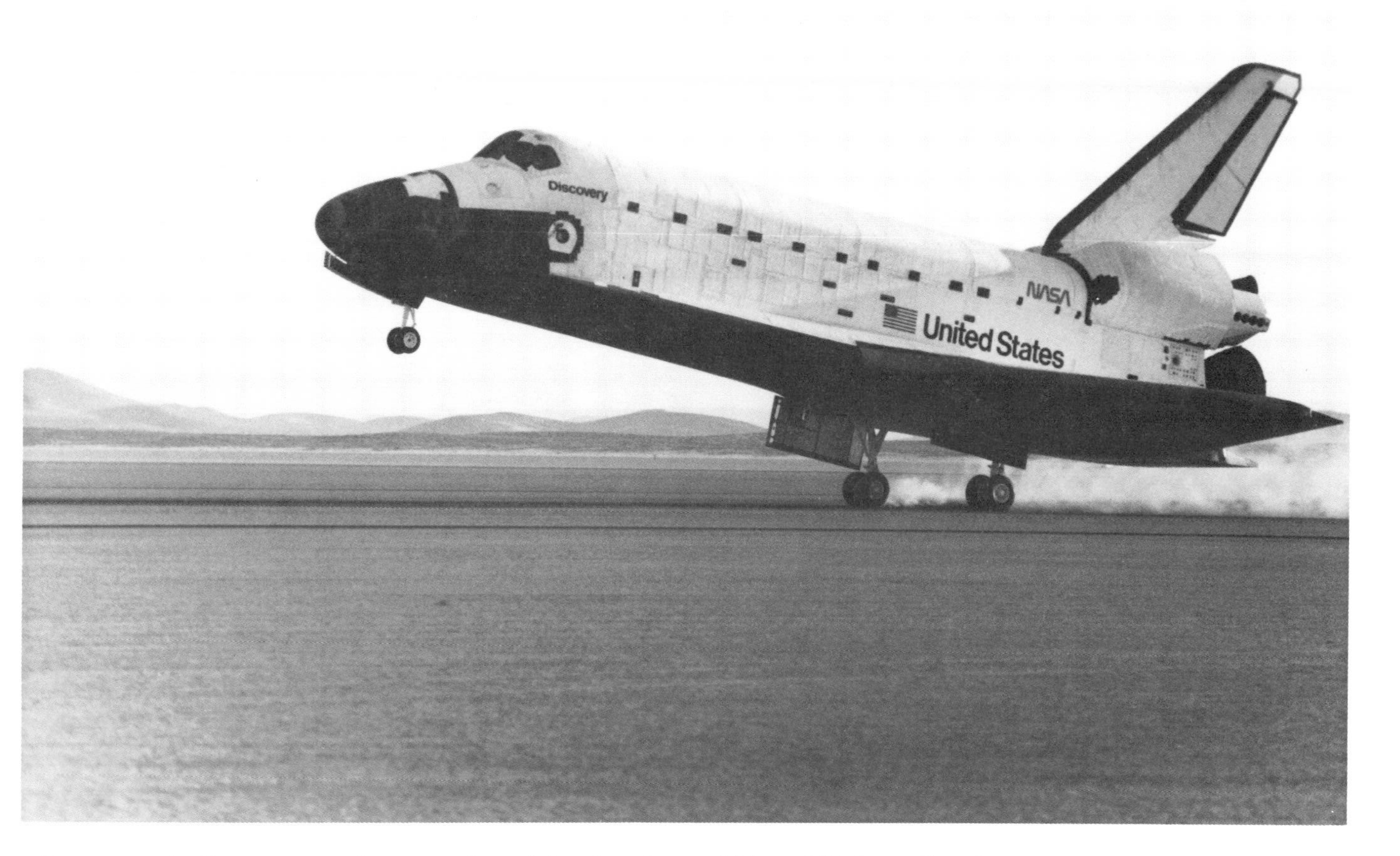

Discovery, with its seven-member crew aboard, touches down at Edward Air Force Base in California on June 24, 1985 (all photos appear courtesy of NASA).

Astronaut Bruce McCandless II performing the first extravehicular activity (EVA) without the use of tethers and umbilicals, just a few meters from the Challenger, on February 7, 1984.

Astronaut Edwin E. Aldrin, Jr. performing the Apollo 11 EVA on the Moon, July 20, 1969.

Mission Control Center's flight control room, Johnson Space Center, Houston, Texas, March 29, 1988.

An artist's concept of the Navstar Satellite in Earth orbit.

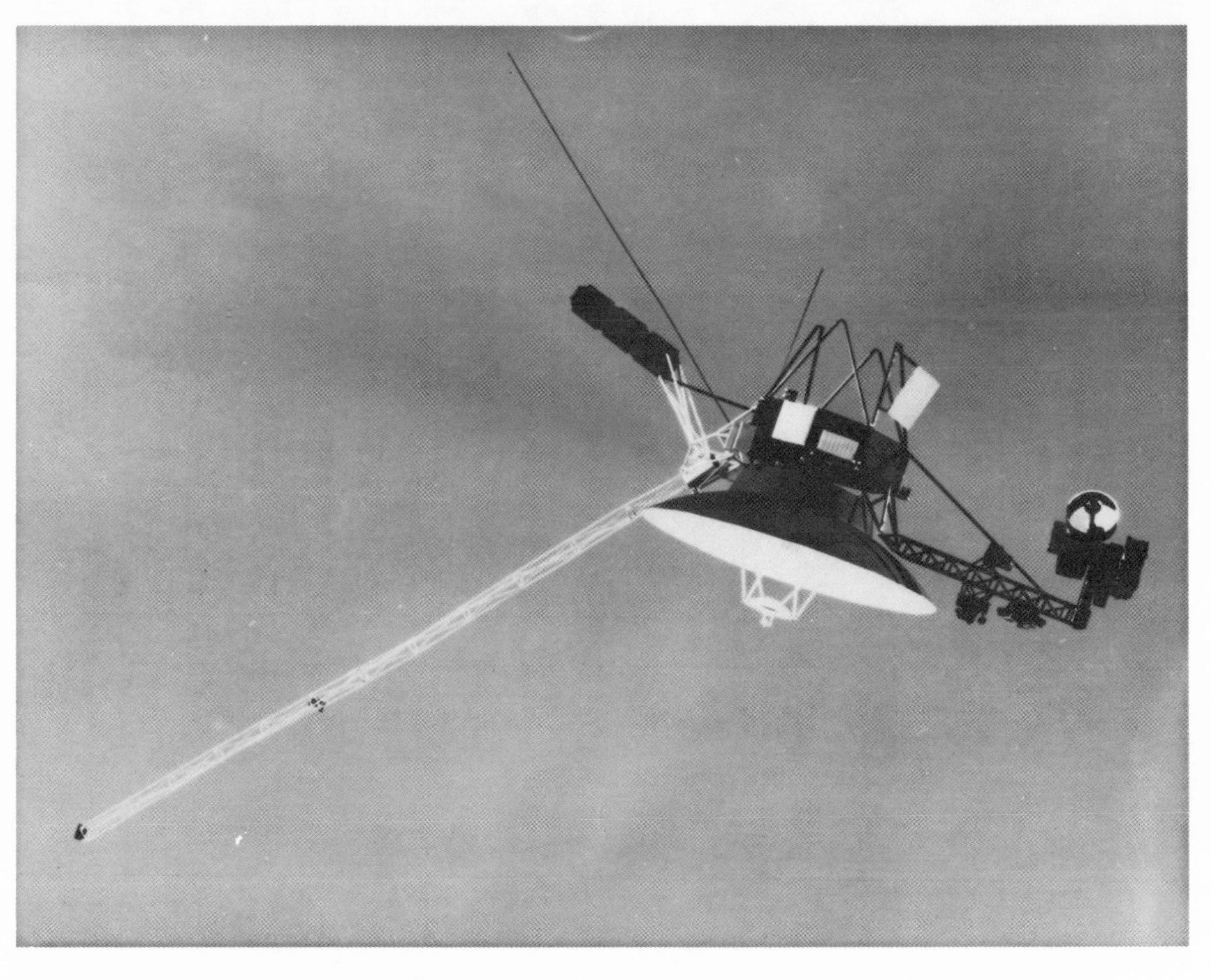

A Voyager spacecraft, full scale model.

Soviet Satellite Cosmos on display at Paris Air Show, June 1975.

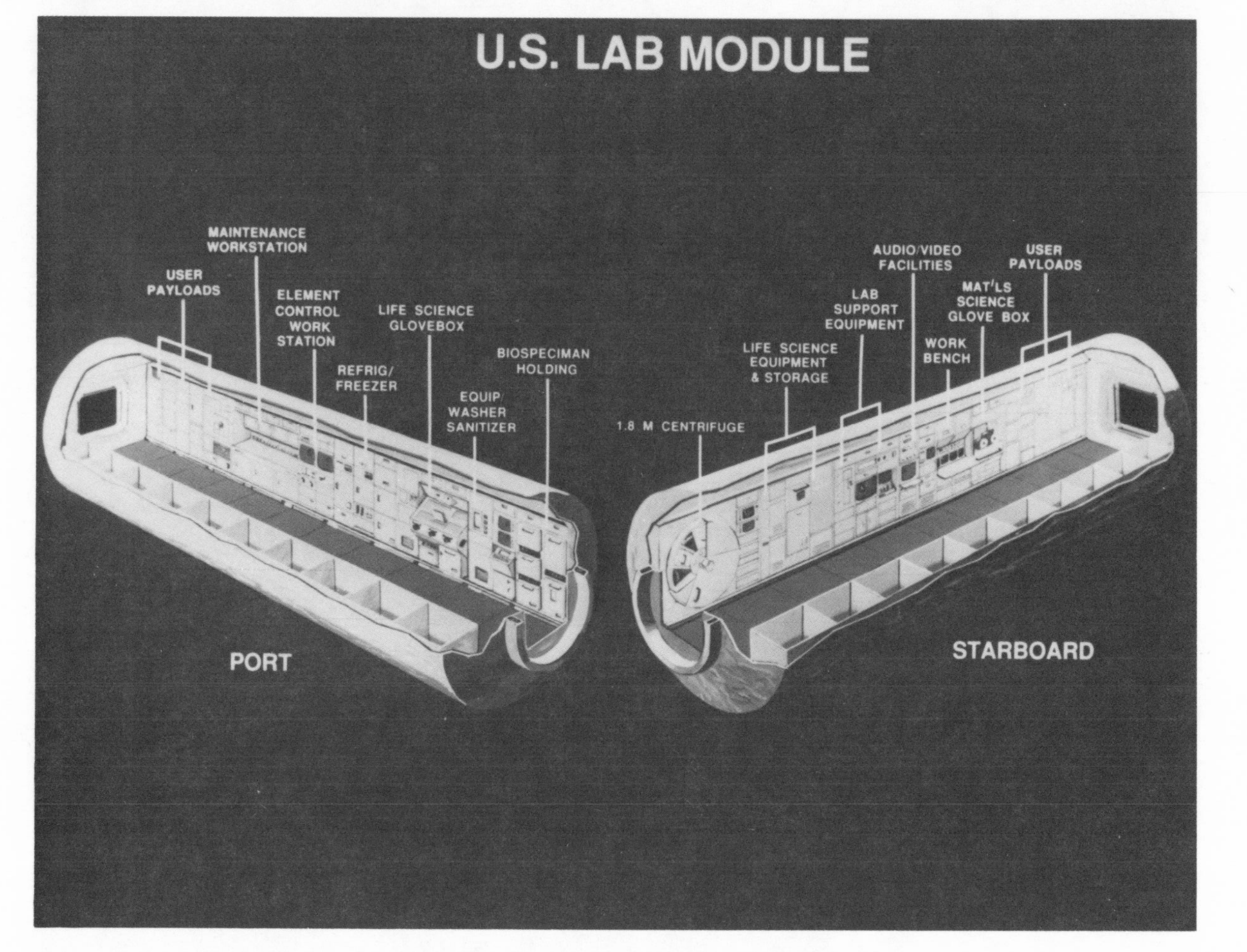

The U.S. Laboratory Module will conduct Life Sciences and Materials research.

The EVA Retriever robot can move about and locate, distinguish, and retrieve objects.

Astronaut Sally K. Ride floating freely on the flight deck of the Challenger, June 21, 1983.

Artist's concept of a possible twenty-first-century space station.

CHAPTER 8

How to Market to NASA

NASA was established by the National Aeronautics and Space Act of 1958 to plan, direct, and conduct aeronautical and space activities for peaceful purposes for the United States. Its goals in space are to enlarge the range of practical applications of space technology and data and to investigate the Earth and its immediate surroundings, the natural bodies in our Solar System, and the origins and physical process of the universe. A more recent goal is using space to help defend the United States.

The purpose of this chapter is to acquaint organizations with how to market their products and services to NASA, whether it be an abstract idea, a manufacturing capability, a fabricated component, construction, basic materials, or a specialized service. In pursuit of this objective, this chapter acquaints prospective contractors with the organizational structure of NASA and briefly describes the major technical program offices and staff offices at the various NASA locations, which provide technical and other assistance, guidelines, and specifications.

WHAT IS MARKETING?

There are several definitions and explanations of marketing, among them the following:

- "Marketing is the process of planning and executing the conception, pricing, promotion, and distribution of ideas, goods, and services to create exchanges that satisfy individual and organizational objectives" (*Marketing News* 1985, p. 1).

- "Marketing is a total system of business activities designed to plan, price, promote, and distribute want-satisfying products and services to present and potential customers" (Stanton 1984, p. 7).
- "Marketing encompasses all activities involving exchange and the cause and effect phenomena associated with it" (Bagozzi 1975, p. 32).
- "Marketing is the process in a society by which the demand structure for economic goods and services is anticipated or enlarged and satisfied through the conception, promotion, exchange, and physical distribution of such goods and services" (Marketing Staff of the Ohio State University 1965, p. 43).
- "Marketing consists of individual and organizational activities that facilitate and expedite satisfying exchange relationships in a dynamic environment through the creation, distribution, promotion, and pricing of goods, services, and ideas" (Pride and Ferrell 1987, p. 7).

Although these definitions of marketing may be acceptable, each has limitations. For example, one of the definitions states that marketing consists of business activities; however marketing also occurs in nonbusiness situations and is practiced by nonbusiness organizations, such as museums and political parties. Another concerns the fact that marketing deals not only with goods and services but also with ideas, causes, and individuals. The last definition appears to be the best of the five, although marketing does not always facilitate "satisfying" exchange relationships; firms sometimes dupe or mislead customers, and some customers do the same to firms. Marketing is also employed in promoting social causes (e.g., family planning) and individuals (e.g., the presidency of the United States and other political offices).

Marketing is practiced by large, medium-sized, and small firms. Some firms are aggressive marketers (e.g., IBM, Coca Cola, Procter & Gamble), while others are less aggressive or passive. Large, wealthy firms tend to be aggressive marketers because of their money and power.

There are a few cardinal rules of effective marketing:

1. Know your market.
2. Offer quality products and services at reasonable prices.
3. Build customer loyalty through rule 2, public relations, and community involvement.
4. Study the competition, and be prepared to carry out retaliatory marketing strategies to maintain customers, profits, and market share.
5. Use the marketing mix variables—product, price, promotion, place or distribution, and public relations—to achieve your objectives in the marketplace.
6. Be familiar with the external environment that affects your business (e.g., the economy, and the legal and political environment.)

Table 8.1 specifies some decisions associated with the marketing mix variables.

KNOW YOUR IMMEDIATE MARKET: NASA

The first cardinal rule of marketing is to know your market. Since NASA buys most of its goods and services from contractors, through the bid process, every contractor and subcontractor doing business with NASA or hoping to do business with it should view NASA as its market. Contractors should become knowledgeable about NASA and how it operates.

NASA's mission is planned, directed, and coordinated from its headquarters in Washington, D.C., the focal point for policy and program formulation. Each of NASA's locations or installations throughout the country, however (e.g., the Johnson Space Center in Houston, Texas, and the Kennedy Space Center in Cape Canaveral, Florida), makes valuable inputs to policies and programs through its respective chief administrator.

Each installation has a specifically prescribed mission, with related tasks, and is allocated the resources necessary for their accomplishment. Although these NASA installations have unique in-house capabilities, their research and operations are pursued mainly through private industry, usually located nearby, with additional support of (nearby) universities and other nonprofit research organizations. For example, the University of Houston at Clear Lake provides some engineering and space commercialization consulting services for the Johnson Space Center. Many corporations nearby, such as IBM (computer software and hardware for Mission Control) and Rockwell International (servicing the U.S. space shuttles), provide important services for the Johnson Space Center. In fact, an important economic phenomenon at every NASA installation is the cadre of new and supporting industries that have mushroomed in close proximity to each installation, bringing thousands of jobs to the respective localities. They include engineering, construction, computer, consulting, printing, and real estate industries. On the downside, when the space industry is in a slump, such as the two years or so following the *Challenger* tragedy, many firms that feed off space contracts awarded by NASA experience economic hardships, and some go out of business. This happened, for example, after the *Challenger* disaster.

OPERATIONS AND NEEDS OF NASA FIELD INSTALLATIONS

Ames Research Center, Moffett Field, California

This center is noted for its technical excellence in life sciences, human factors and human-machine interactions, fluid dynamics and heat transfer, aerodynamics and flight dynamics, flight stability and control, and technical project management. As the NASA center with primary responsibility for

Table 8.1
Decisions and Activities Associated with Marketing Mix Variables

Marketing Mix Variables	Decisions and Activities
Product/Service	Develop and test-market new products; modify existing products; eliminate products that do not satisfy customers' desires; determine sizes, performance levels, quality standards; formulate brand names and branding policies; create product warranties and establish procedures for fulfilling warranties; plan packages including materials, sizes, shapes, colors, and designs; develop quality services, with on-time delivery.
Price	Analyze competitors' prices; formulate pricing policies; determine method or methods used to set prices; do break-even and profitability analyses; set prices; determine discounts for various types of buyers; establish conditions and terms of sales; monitor prices and revenues; change prices if necessary.
Distribution	Analyze various types of distribution channels; design appropriate distribution channels; design an effective program for dealer relations; establish distribution centers; formulate and implement procedures for efficient product handling; set up inventory controls; analyze transportation methods; minimize total distribution costs; analyze possible locations for plants and wholesale or retail outlets; establish security system for distribution of sensitive products.

Table 8.1 (Continued)

Marketing Mix Variables	Decisions and Activities
Advertising & Promotion	Establish market target(s), sizes, locations, business profile; set promotional objectives; determine major types of promotion to be used; select and schedule advertising media; develop advertising messages; measure the effectiveness of advertisements; recruit and train salespersons; formulate compensation programs for sales personnel; establish sales territories; plan and implement sales promotion efforts such as free samples, coupons, displays, sweepstakes, sales contests, and cooperative advertising programs.
Public Relations	Prepare and disseminate public service announcements, press kits, news stories about the company, its people, and products; establish PR office if necessary.

research in the life sciences, the Ames life sciences program covers many disciplines and research areas. Biomedical research includes developing and operating hardware and experiments for determining the effects of space flight on humans and nonhuman living organisms and for providing information to solve space medicine problems. Advanced life support system research is conducted to develop techniques for sustaining human life and maintaining human efficiency in space. Extraterrestrial life detection is concerned with the origin of life and with the abundance and distribution of life and life-related compounds in space and on other planets.

Ames is also noted for its research work in aeronautics, astronautics, and space sciences. Currently the areas of greatest interest in aeronautics are in development of vertical or short takeoff and landing (V/STOL) aircraft for military aircraft applications and urban regional transportation systems. In astronautics, research in space sciences, Earth applications, and spacecraft development are emphasized. There is also interest in associated technology development in support of astronautics programs, such as infrared sensors, cryogenics, atmosphere entry aerothermodynamics and thermal protection systems, and computational chemistry.

Space sciences research includes astrophysics, astronomy, studies of planetary atmosphere, and climate and stratosphere research. To this end, a great deal of emphasis is being placed on spacecraft development, remote sensing technology, space telescopes, and so on.

Ames Research Center and Dryden Flight Research Facility, Edwards Air Force Base, California

The mission of the Dryden Flight Research Facility (DFRF) is the conduct of research on flight and the problems of manned flight within the atmosphere. This research involves work on problems of takeoff and landing, low-speed flight, supersonic and hypersonic flight, and re-entry into the Earth's atmosphere. Some of the research is done with remotely piloted aircraft. The small business specialist at Ames Research Center is available to help firms in identifying potential procurement opportunities at both Ames and DFRF and encourages inquiries.

Goddard Space Flight Center, Greenbelt, Maryland

Goddard Space Flight Center (GSFC) is located 10 miles northeast of Washington, D.C., on about 1,100 acres in Greenbelt, Maryland. It is staffed by more than 3,500 government employees, including some of the world's leading scientists, engineers, and technicians, and some 2,000 contractor personnel.

GSFC's prime responsibility is the development and management of application satellite projects, unmanned scientific satellite projects, and worldwide NASA tracking and data acquisition operations. The center's scientific staff is concerned primarily with research into magnetic fields, energetic particles, ionospheres and radio physics, planetary atmospheres, interplanetary matter, solar physics, communication, astronomy, and meteorology.

Examples of satellite technology utilization having development roots at Goddard are the (1) *Landsat* satellite, which scans the Earth's surface every eighteen days gathering a wide range of Earth resources survey information; (2) the Geostationary Operational Environmental Satellite (GOES), which provides timely global weather information via black and white, television-like images; (3) *Intelsat IV* satellites, which are geostationary over the Atlantic, Pacific, and Indian oceans and serve more than 106 nations around the world; (4) Canada's *ANIK I* and *II*; and (5) Western Union's *WESTAR A* and *B* satellites.

GSFC has a number of major missions. One is the development of a satellite tracking system, Tracking and Data Relay Satellite System (TDRSS), which places the satellite tracking station in space in geosynchronous orbit and thus, in effect, looks down and at orbiting satellites. The TDRSS will permit greatly expanded satellite coverage and will be capable of handling the high data rates of future missions such as *Spacelab* and *Landsat D*.

Another major mission at GSFC is project management of the *Delta* launch vehicle, which has placed into orbit more than 150 successful unmanned satellites for NASA, DOD, domestic communications corporations, and numerous foreign countries. Goddard is the home of the National Space Science Data Center, the central repository for the scientific data collected from space science experiments.

In October 1981, GSFC was consolidated with the Wallops Flight Facility, Virginia. The Wallops launch range provides vehicle assembly and launch facilities, communications, tracking instrumentation, data acquisition and data processing of sounding rockets, re-entry vehicles, and balloons and satellites launched from Wallops Island and other off-site locations.

In order to maintain an active support of small and disadvantaged businesses, Goddard maintains an automated source system comprised of about 4,000 firms. The center encourages businesses to submit a Bidders Mailing List Application (SF 129) to be added to the Goddard Automated Source System and to contact the small business specialist for advice on procurement procedures and opportunities.

Lyndon B. Johnson Space Center, Houston, Texas

The Johnson Space Center (JSC) is a focal point of the nation's manned spaceflight activities, including spacecraft development, program management, crew training, space flight operations, and related medical research and life sciences. JSC is also responsible for conducting investigations of lunar science, space science, and Earth resources technology and applications. Major programs that have been assigned to JSC include Mercury, Gemini, Apollo, Skylab, Apollo/Soyuz, the space shuttle, Earth Resources, Space and Life Sciences, and the development of the U.S./International Space Station.

All of JSC's programs require tremendous amounts of materials and services, which must be obtained from outside the government. Material needs range from raw materials and commercial items to sophisticated spacecraft; services range from housekeeping to engineering, medical, and scientific capabilities. The small business specialist at JSC serves as a focal point to assist companies in understanding the center's needs.

Continuous product and service requirements exist to support many programs, including the following:

1. The space shuttle orbiter, a reusable space airplane that carries satellites and scientific payloads into orbit.
2. *Spacelab,* a joint venture between NASA and ESA to produce and operate in space a reusable laboratory that will be available to an international community of users in applied sciences, life sciences, and advanced technology.
3. NASA's Earth Resources Program for improved identification and use of mineral and land resources, mapping and charting, urban land use, and agricultural and forestry resources.
4. The Space and Life Sciences Program, which includes life sciences, medical research, lunar and planetary science, space sciences, and biomedical research on the physiological stress of space flight on man.

John F. Kennedy Space Center, Kennedy Space Center, Florida

The Kennedy Space Center (KSC) is the major NASA launch organization for both manned and expendable (unmanned) space missions. KSC launched the *Apollo* and *Skylab* space vehicles and the U.S. space shuttles, and is a primary landing site for space shuttle orbiters upon completion of their missions. The center also launches a wide variety of expendable spacecraft, including scientific probes of the far reaches of the Solar System. A host of technical and administrative activities support KSC missions, such as design engineering, testing, assembly and checkout of launch vehicles and payloads, and associated purchasing and contracting.

Langley Research Center, Hampton, Virginia

The Langley Research Center conducts extensive research in aeronautics, space technology, electronics, and structures. Aeronautics has been a Langley specialty for more than sixty-five years. Research in all aircraft speed ranges, from subsonic to hypersonic (ten times the speed of sound), accounts for about two-thirds of Langley's work. Specific programs concern improvements in the efficiency of transport aircraft and research on transonic transports and transonic and supersonic military aircraft.

Space technology research at Langley places emphasis on support of NASA's space shuttles and their payloads. Langley's structures work is directed toward research in materials, structures, and loads. Composite materials that can reduce weight in aircraft and space shuttle structures are of special interest, as are thermal protection materials.

Lewis Research Center, Cleveland, Ohio

The Lewis Research Center directs a great deal of its activities toward advancing technologies for aircraft propulsion, propulsion and power generation for space flight, space communications systems, and new terrestrial energy systems and automotive engines. Specific projects include development of engines that will operate as quietly, cleanly, and efficiently as possible; studies of alternate fuels for jet aircraft; and basic and applied research on materials and metallurgy, combustion processes, seals, bearings, gears and lubrication, and system control dynamics. The diverse nature of the research activities at Lewis Research Center offers many procurement opportunities for small and disadvantaged firms. These companies should seek the assistance of the Lewis small business specialist.

Marshall Space Flight Center, Huntsville, Alabama

The Marshall Space Flight Center (MSFC) is one of the primary NASA centers for the design and development of space transportation systems, orbital systems, scientific and applications payload, and other systems for present and future space exploration. Principal MSFC roles include rocket propulsion systems; design and development of manned vehicle systems; Spacelab mission management and payload definition; design and development of large and specialized automated spacecraft; and management of space processing activities.

Examples of a few major programs at MSFC are:

1. Space shuttle: MSFC is responsible for the design, development, testing, and evaluation of the space shuttle main engine, the solid rocket booster, the solid rocket motor, and the external tank.
2. *Spacelab:* MSFC has program responsibility for this major international cooperative program between NASA and ESA. *Spacelab* is designed as a versatile and scientific modular laboratory to be carried in the shuttle orbiter bay.
3. Materials processing in space: MSFC is responsible within NASA for exploring and developing the potential for MPS. Growth of superior single crystals for solid-state electronics, high-strength materials, and separation of living cells for pharmaceutical products all show great promise.
4. Space telescope: NASA's space telescope (ST) is a multipurpose optical telescope that, when placed in Earth's orbit, will enable scientists to observe objects in space much more clearly than through the best telescopes on Earth.
5. Flight experiments: MSFC conducts many flight experiments to predict and track severe storms by satellites; to determine the effects of long duration exposure of materials and equipment to a space environment; and to study gamma ray bursts and transients.
6. Studies of future space systems: Among the many future space systems MSFC is studying are: the U.S./International Space Station; unmanned space platforms

for the conduct of space science experiments; an advanced X-ray astrophysics facility, which would be a national facility for the study of X-ray sources; and a Gravity Probe-B satellite, which will test Einstein's theory of relativity.

Since MSFC is one of the largest NASA centers, it provides many contracting and subcontracting opportunities. Companies interested in MSFC procurement programs should contact the MSFC small business specialist for assistance.

National Space Technology Laboratories, NSTL Station, Mississippi

The National Space Technology Laboratories (NSTL) is the prime NASA installation for static test firing of large rocket engines and propulsion systems. NSTL is heavily involved in support of the shuttle test program (e.g., space shuttle main engine testing) and is conducting research in terrestrial applications. In terrestrial applications programs the focus of the installation's capability is in its Earth Resources Laboratory (ERL), currently engaged in remote-sensing technology research and development.

Jet Propulsion Laboratory, Pasadena, California

The Jet Propulsion Laboratory (JPL) of the California Institute of Technology is a government-owned research, development, and flight center that performs a variety of tasks for NASA. JPL works under close direction from NASA headquarters, with day-to-day administration and coordination provided by the NASA Resident Office.

JPL's primary role is the scientific investigation of the planets and deep space using automated spacecraft. In addition, JPL conducts projects to develop and apply new technologies to the solution of problems on Earth, with emphasis on solar energy and conservation studies.

JPL's current tasks include the Voyager mission to the outer Solar System and the planets Uranus and Neptune, the continued exploration of Jupiter with the *Galileo Orbiter* and probe, and the U.S. management of the infrared astronomical satellite and the International Solar Polar Mission. Supporting research and advanced development are conducted in electric propulsion, aerothermodynamics, fluid physics and electrophysics, applied mathematics, space power generation, planetary atmospheres, long-range communications, and systems simulation and analysis techniques.

Because of the heavy emphasis on research and development, JPL seldom buys in large quantities, with the exception of electronic components. A significant portion of JPL's procurement budget is spent on the following products and services: fabrication, electronic components, electronic instrumentation and test equipment, and miscellaneous supplies and services.

The JPL small business specialist or the minority business specialist will assist companies in identifying procurement opportunities at JPL.

THE PROCUREMENT PROCESS

The procurement process (Figure 8.1) typically is initiated at a NASA installation when a program or project office there submits a procurement request to the procurement office. The assigned contracting officer, in consultation with the small business representative and the technical officer, will then make several key decisions:

1. Whether the required supplies or services are available from other government sources, such as stock items at a General Services Administration (GSA) supply depot. If so, the contracting officer must acquire the items from the depot directly or from suppliers on the federal supply schedule. (Firms interested in getting their name and products listed on the federal supply schedule should contact a Business Service Center of the GSA for additional information.)
2. Whether the procurement should proceed under one of the special assistance programs, such as setting the procurement aside for the exclusive participation of small or minority business.
3. The appropriate method of conducting the procurement: sealed bids, competitive proposals, or other competitive procedures, or a noncompetitive bid when there is only one responsible or reliable source.

Once these decisions are made, the contracting office prepares the solicitation package and arranges for the announcement of the solicitation through various channels.

Announcement of the Solicitation

NASA installations use several methods to inform prospective bidders of contracting opportunities. First, each installation maintains its own source file of information regarding the capabilities and products and services of companies expressing an interest in doing business with it. When a procurement need arises, NASA prepares a bidders' list based on information in its files and other available information. It then sends a copy of the solicitation to enough companies on the list to ensure adequate competition. Some of the criteria NASA uses to select companies for the bidders' list are past performance, capability, reputation, satisfaction with previous work done for NASA, and length of time in business.

A second method NASA uses to announce solicitations is through advertisements in the *Commerce Business Daily (CBD)*, the official announcement medium for federal procurements. A notice of the pending procurement is normally placed in the *CBD* at least fifteen days prior to

Figure 8.1
NASA Procurement Process

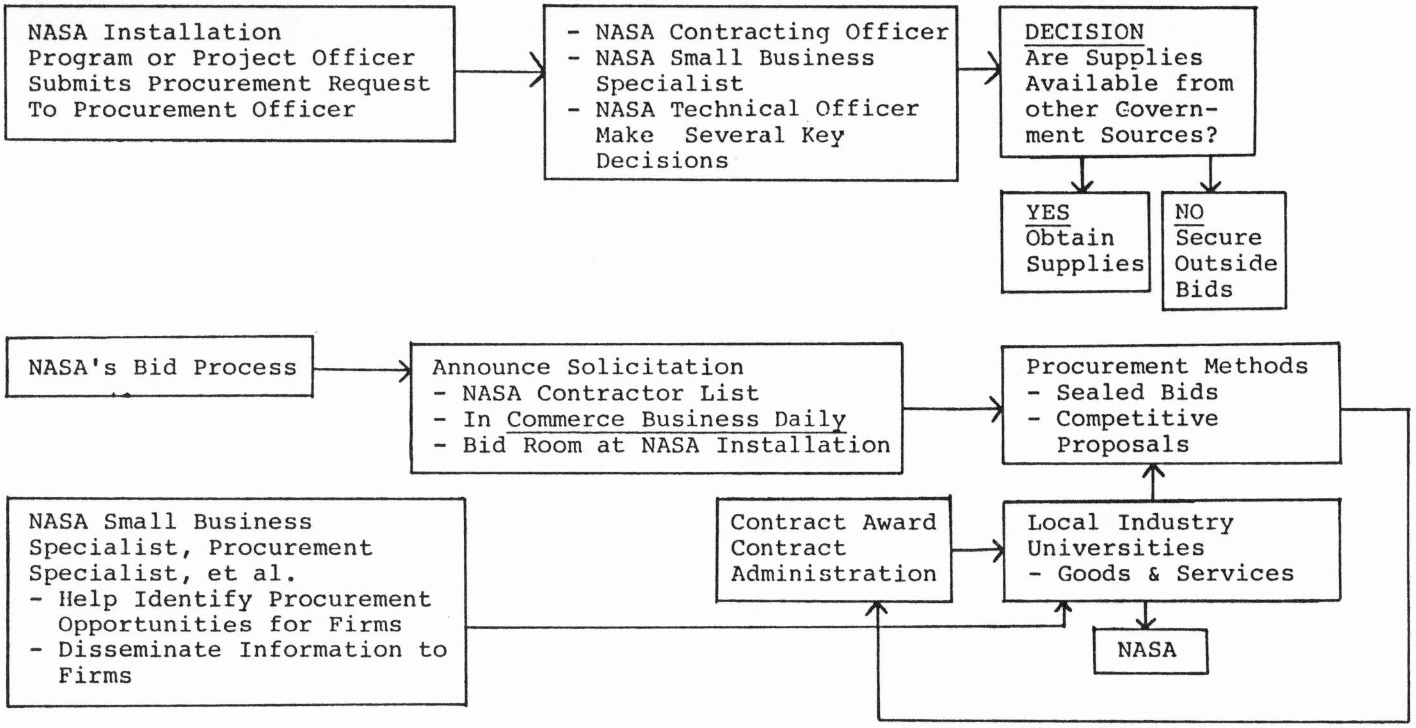

NASA Installation
Program or Project Officer
Submits Procurement Request
To Procurement Officer

- NASA Contracting Officer
- NASA Small Business Specialist
- NASA Technical Officer Make Several Key Decisions

DECISION
Are Supplies Available from other Government Sources?

YES
Obtain Supplies

NO
Secure Outside Bids

NASA's Bid Process

Announce Solicitation
- NASA Contractor List
- In Commerce Business Daily
- Bid Room at NASA Installation

Procurement Methods
- Sealed Bids
- Competitive Proposals

NASA Small Business Specialist, Procurement Specialist, et al.
- Help Identify Procurement Opportunities for Firms
- Disseminate Information to Firms

Contract Award
Contract
Administration

Local Industry
Universities
- Goods & Services

NASA

issuance of the solicitation when the anticipated value of the contract exceeds $10,000. A copy of the solicitation is also sent to companies requesting it on a first-come, first-served basis. A subscription to *CBD* is worthwhile in order to keep abreast of solicitation announcements and information on contract awards and other government business announcements. A subscription may be obtained from the Superintendent of Documents, Government Printing Office, Washington, D.C. 20402.

A third source of information on NASA procurements is the bid room at each NASA installation, which has on file current solicitations available for inspection by prospective contractors. A central bid room with information on all open NASA solicitations is maintained by the Headquarters Small Business Office in Washington, D.C.

Procurement Methods

There are two main procurement methods: sealed bids and competitive proposals. The key difference is that the main criterion for evaluating a sealed bid is price, whereas competitive proposals allow for consideration of factors other than price, such as technical capabilities of the prospective contractor, anticipated performance, delivery time, and reputation of the contractor.

In the sealed bid procurement method, the specifications or exact performance requirements can usually be clearly stated, required delivery dates are known, and the contract is simply awarded to the lowest bidder. A solicitation under this procurement method is called an Invitation for Bids (IFB), and the response is a bid. A solicitation for competitive proposals is called a Request for Proposals (RFP), and the response is a proposal.

Although sealed bidding is the preferred method of awarding contracts, NASA's heavy emphasis on research and development normally requires a more flexible contracting approach. Consequently most NASA procurements are accomplished through competitive proposals. With competitive proposals the selection process generally has four steps:

1. Evaluation of proposals to determine which offers are the best ones.
2. Negotiation or discussions with these firms.
3. Tentative selection of the apparently successful firm.
4. Final negotiation with that firm prior to award of the contract.

Evaluation of proposals is often performed by a Source Evaluation Board, which consists of a group of NASA officials familiar with the two procurement procedures and the technical requirements of the RFP. A detailed description of this evaluation-selection process is contained in the NASA *Source Evaluation Board Manual,* available from the Superintendent of Documents, Government Printing Office, Washington, D.C. 20402.

Contracting Considerations

Prospective contractors must meet or agree to meet many requirements prior to being awarded a NASA contract. Some of the most important ones are in the areas of bonding, qualifications of the contractor, equal opportunity, reliability and quality assurance, safety and health, industrial relations and security. For example, under the provisions of the Miller Act, NASA construction contractors are required to post performance and payment bonds on contracts in excess of $25,000 to protect the government's interest in the proper and timely completion of the work and to secure payment for labor and material furnished under the contract.

The Architect-Engineer Selection Board at each NASA field installation reviews the qualifications of firms interested in performing architectural and engineering work in connection with NASA construction projects. In the case of procurements estimated to cost more than $10,000, the board also conducts oral or written discussions with a minimum of three firms. It then submits a report to the NASA installation director recommending, in order of preference, those firms considered best qualified to perform the services required. Upon approval by the installation director of the list of qualified firms, contract negotiations are conducted with the first-ranked firm. If a mutually satisfactory contract cannot be agreed to, negotiations are then initiated with the second firm on the list. This procedure continues until a contract has been negotiated.

Each Architect-Engineer Selection Board maintains a list of qualified firms for various types of projects. Firms interested in NASA projects should file Standard Form 254 (Architect-Engineer and Related Services Questionnaire) with the various NASA field installations and with NASA Headquarters, Office of Facilities. Firms should keep their qualification information as current as possible. On special projects, such firms may also be required to file Form 255 (Architect-Engineer and Related Services Questionnaire for Specific Project).

Each NASA installation vigorously pursues additional contracting guidelines and considerations, such as equal opportunity, reliability and quality assurance, safety and health programs, and security clearance. Contractors or subcontractors are obligated not to discriminate against any employee or applicant because of age, religion, race, color, sex, or national origin, and is required to take affirmative action to ensure equal employment opportunity. Many NASA publications spell out requirements relative to reliability and quality assurance. There are also specific provisions in each contract to help avoid loss of life, personnel injury or illness, and loss of property. Occupational Safety and Health Act standards are invoked, as well as other federal and industry standards.

Should a NASA contract or solicitation require access to classified information, an industrial security clearance will be required. This occurs frequently with Department of Defense (DOD) contracts. It often involves FBI, CIA, credit, and general background checks on the firm and its employees, Further procedures and requirements are found in the *Industrial Security Manual for Safeguarding Classified Information*. This document, like many others, can be purchased from the Superintendent of Documents, U.S. Government Printing Office, Washington, D.C. 20402.

Contract Administration

Once NASA awards a contract and the performance begins, the government initiates the process of contract administration. This involves accounting, reporting, and other contractual requirements associated with performing on government contracts. (Details of contract administration are beyond the scope of this chapter; they may be obtained from the NASA installation.) Contract administration is the responsibility of the NASA contracting officer, but certain functions may be delegated by interagency agreement to cognizant agencies of DOD (such as the Defense Contract Administration Service and the Defense Contract Audit Agency).

MARKETING YOUR CAPABILITIES

The U.S. government is the largest purchaser of goods and services in the world. Collectively, government agencies spend approximately $160 billion annually. Despite this enormous expenditure, however, government contracts are not easily obtained; competition is fierce. This section of the chapter is designed to be a primer on how to market to NASA. Although the information presented here is specific to NASA, many of the principles are applicable to marketing to other federal, state, and local agencies. Figure 8.2 illustrates the major steps in a firm's marketing efforts to NASA.

Understanding the Market

The first step in marketing is understanding the market: NASA, which encompasses its field installations. Each installation has several specialists, such as procurement specialists and the small business specialist, who can provide useful guidance regarding NASA guidelines and the firm's prospects of doing business with that NASA installation. Each NASA installation provides useful publications regarding procurement guidelines, major contractors, contract guidelines, and so on.

Total NASA procurements with business firms totaled $9 billion in FY 1988, with thousands of companies participating in the process. After becoming familiar with the predominant types of goods and services NASA buys, the locations where they are bought, and the aggregate dollar amount of purchases in a given area, the firm must give careful consideration to the

Figure 8.2
Marketing to NASA

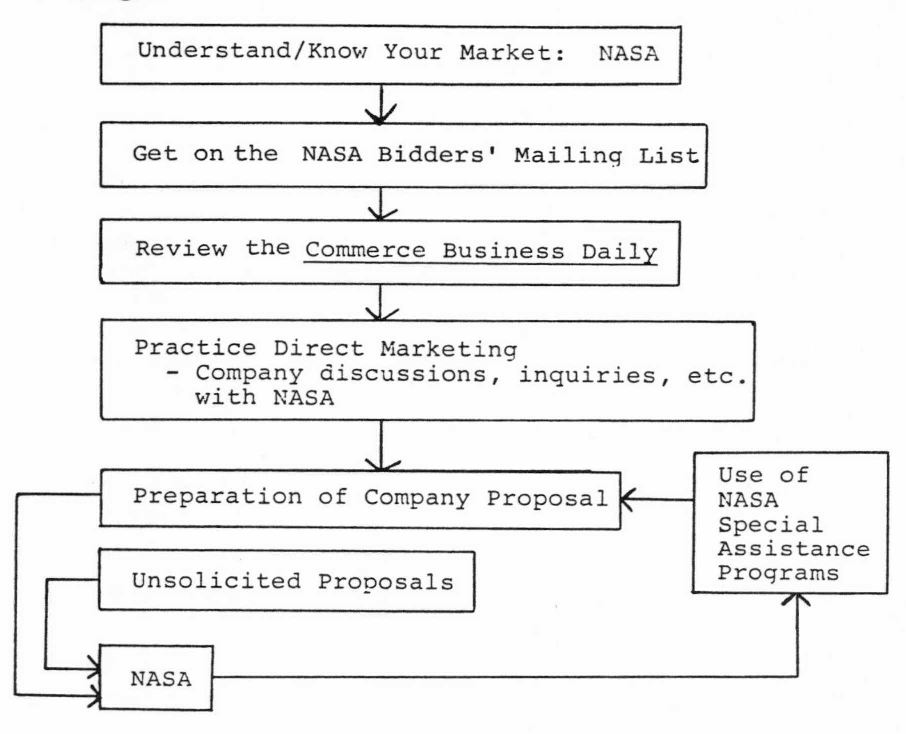

extent to which NASA represents a viable market—that is, how close the match is between the firm's capabilities and NASA's needs. It is important not to expend too much marketing effort in instances in which the prospect of a contract does not appear likely.

Once a company determines that NASA does represent a viable market, there are several approaches the company can pursue simultaneously to increase its likelihood of finding specific procurement opportunities.

Finding Specific Opportunities

The Bidders' List. NASA uses an internal bidders' mailing list to identify potential bidders for a procurement. Since every NASA installation maintains its own list, it is essential to apply individually to each installation at which a company hopes to do business. The application consists of a Standard Form 129 requiring information about the company, as well as specific information regarding its products and services. There are special forms for architecture and engineering firms. Prospective contractors should also submit bidders' mailing list applications to other government agencies active in their field of interest.

Commerce Business Daily. The *Commerce Business Daily (CBD)* lists federal government projects out for bid, bid requirements, winners of contracts, and so on. Firms that are serious about obtaining government business ought to review *CBD* regularly to keep abreast of government contracts, identify interesting solicitations as they are issued, and request copies of those on which they would like to bid. Generally all procurements over $10,000 (or $25,000 for construction) are described in *CBD*. Subscriptions may be obtained from the Superintendent of Documents, Government Printing Office, Washington, D.C. 20402. *CBD* is also available in most public libraries.

Direct Marketing. One of the most productive techniques for identifying procurement opportunities within NASA is by direct contact with technical requirements personnel. NASA's technical and procurement personnel welcome inquiries and discussions with appropriate company representatives. The Small Business Office at each center arranges visitations and responds to written inquiry.

Large, well-known corporations, such as IBM, Rockwell International, and Ford Aerospace, place heavy emphasis on direct marketing to obtain NASA contracts. They have spent millions of dollars to establish large office complexes near NASA centers (sometimes literally across the street from NASA, as in the case of the Johnson Space Center). Such close proximity provides obvious advantages, such as ease of meeting with NASA officials, expeditious arrangement of top-level meetings with NASA officials to help solve technical and other problems, and establishment of long-term rapport with NASA officials. Some of the leading scientists and engineers at companies like IBM also have offices within NASA centers from which they coordinate NASA-company business. Some companies hire lobbyists who work out of Washington, D.C.

This is not meant to imply that direct marketing alone will secure a NASA contract for a firm. It suggests that direct marketing is essential to securing NASA contracts. Other company requirements include a good track record, expertise, and reliability.

Preparation of Proposals

The preparation of an effective proposal is perhaps the most important part of the marketing process. Details of an effective proposal are beyond the scope of this chapter; however, it is necessary to mention that effective proposals are well written, clear, detailed, follow exact specifications, and are submitted on time. As a general rule, companies that are not familiar with preparing proposals for government contracts should seek professional assistance. University or private consultants, and in some cases local business assistance centers, are possible sources of such assistance. Many NASA publications provide important information on technical specifications and contract guidelines. Two such sources, in addition to those available at the local NASA center, are the *Index of Federal Specifications and Standards* and the *Department of Defense Index of Specifications and Standards.*

Unsolicited Proposals

An unsolicited proposal is a written offer to perform a task or effort, prepared and submitted by an organization on its own initiative without a formal solicitation from NASA. Although most NASA R&D contracts are awarded as a result of standard competitive procurement procedures, another important method of doing business with NASA is through the submission of relevant new ideas and concepts in the form of unsolicited proposals. In general, research of a fundamental nature—that which bears potential for advancing the state of the art in a particular area, contributes to knowledge of a specific phenomenon, or provides fundamental advances in engineering or the sciences—is appropriate for the unsolicited approach.

Guidelines for the preparation and submission of unsolicited proposals may be obtained by writing to the small business specialist at any NASA installation. Another source of information on NASA programs is through the issuance of periodic notices in the form of "Dear Colleagues" letters, which disseminate information to members of the scientific and engineering community. Firms should request that they be placed on the mailing for such letters.

SPECIAL ASSISTANCE PROGRAMS

NASA also has special assistance programs that serve as useful adjuncts to a firm's program to market to NASA. These programs are designed to ensure that small or minority-owned businesses have a fair opportunity to participate in federal procurement. Some of these programs include small business set-asides and programs for minority-owned companies.

Small Business Set-Asides

Procurements under $10,000 that are subject to small purchase procedures are reserved exclusively for small business. Additionally, certain classes of acquisitions are frequently set aside for bidding by small concerns. In some instances, portions of large procurements may be set aside for exclusive small business bidding. The *CBD* contains announcements about set-asides.

Minority Business Enterprise

NASA's Minority Business Enterprise Program is directed toward ensuring the equitable participation of minority firms in NASA's prime and subcontract opportunities. NASA works closely with the SBA to ensure that minority firms get fair opportunities to bid on NASA contracts. These minority firms are usually owned by (but not limited to) black Americans, Hispanic Americans, native Americans, Asian-Americans, and other minority groups. Qualifying firms interested in participating in this program should contact the nearest SBA office, in addition to making their capabilities known to NASA.

Labor-Surplus-Area Set-Asides

Under labor-surplus-area set-asides, competition for contracts is restricted to firms with production facilities in labor-surplus areas (areas of higher than average unemployment) even if their headquarters is not located in such areas. The firms must agree to perform most of the contract work in the labor-surplus areas. The U.S. Department of Labor defines and classifies labor-surplus areas. For current information on qualifying areas, consult *Area Trends in Employment and Unemployment*, available from the Superintendent of Documents, Government Printing Office, Washington, D.C. 20402.

Women-Owned Businesses

NASA makes special efforts to advise women about business opportunities and preferential contracting programs for which they may be eligible. For example, it reviews bidders' mailing lists to ensure a fair representation of women-owned firms and holds special conferences to assist women-owned or women-controlled businesses in the process of doing business with NASA.

Subcontracting Opportunities for Small Business

Subcontracting with NASA prime contractors is an important source of revenue for many companies, both large and small. Companies can seek out subcontracting opportunities by identifying their capabilities to NASA's major prime contractors (for example, General Electric, McDonnell Douglas, and Rockwell International). Other useful sources of information to assist small businesses in determining appropriate prime contractors to contact are the *CBD*, directories of major government contractors published by the Small Business Administration and the Department of Defense, NASA's *Annual Procurement Report*, which contains a list of the top 100 contractors ranked according to the amount of NASA contract awards, and the small business specialist at each NASA installation.

Procurement Conferences

NASA is an active participant in the Federal Procurement Conference Program sponsored by the Department of Commerce predominantly to assist small business. At these conferences, NASA representatives counsel many of the participants on a one-on-one basis and follow up with many small businesses to assist them in marketing their products and services to NASA. NASA small business specialists at all of the field centers also work closely with local chambers of commerce and trade associations to advise them of NASA's needs and to aid interested firms in informing NASA of their capabilities and expertise.

Other Forms of Assistance

Among the many other forms of assistance to help businesses market their products and services to NASA are the following:

- *NASA Tech Briefs,* a monthly journal containing articles on innovations and improved products or processes developed for NASA that are thought to have commercial value.
- The Computer Software Management and Information Center (COSMIC), located at the University of Georgia, Athens, which collects all of the computer programs NASA has developed (and also some of the best computer programs developed by other government agencies), verifies that they work properly, and makes them available at reasonable prices.
- Industrial Applications Centers (IACs), established by NASA to assist small business and the nonaerospace industrial sector in making profitable use of new knowledge that has resulted from aerospace research and development. Each IAC is based at a university or a not-for-profit research institute and is staffed with specialists skilled in the use of computer search and retrieval techniques. The IAC charges a modest fee for a wide variety of services offered to IAC clients.

SUMMARY

This chapter discusses basic marketing steps that firms should follow if they wish to sell products and services to NASA.

Marketing is described as an important functional area of business and nonbusiness activity that involves product and service development, pricing, distribution, promotion, packaging, and public relations to achieve organizational objectives, which may include profit, market share, and customer satisfaction. There are many cardinal rules of marketing, such as knowing the market and the competition and striving for customer satisfaction.

Firms interested in marketing goods and services to NASA can determine NASA's operations and needs through contacts and consultations with NASA. The firm should also become acquainted with NASA's procurement process through such vehicles as NASA's solicitation announcements, the *CBD*, and contacts with NASA's procurement specialists.

There are many important steps in marketing the firm's capabilities to NASA: getting on the NASA bidders' list, direct marketing, and proposal preparation and presentation. In its marketing effort, the firm should take advantage of special assistance programs at NASA, such as small business set-asides, the Minority Business Enterprise Program, and special opportunities for women-owned business. Other assistance can be derived through procurement conferences, Industrial Applications Centers, and *NASA Tech Briefs*.

There are profits to be made in space business, but marketing know-how is key to such profitability. Kotler (1988), Bennett (1988), Pride and Ferrell (1987), Stanton and Futrell (1987), McCarthy and Perreault (1987), Bagozzi (1986), Dalrymple and Parsons (1986), and Enis and Cox (1985) are excellent books for further reading on basic marketing.

CHAPTER 9

Space Insurance: Its Role and Importance

PARTICIPANTS

Figure 9.1 is a diagram of participants in the multistaged process of insuring spacecraft (satellites). The first participant involved is the owner/user. These individuals or companies have a need for satellite information, so they purchase the equipment from the manufacturers. Examples of owners/users are Radio Corporation of America (RCA), the government of Indonesia, and Western Union and Telegraph Company.

The second participant is the *broker,* whose role is that of a middleman. Brokers arrange for coverage that is bought by the owner/user. Prospective brokers negotiate the amount and type of coverage for the owners and then secure the coverage in the marketplace. Examples of brokers in the United States are Corroon & Black/Inspace, and Marsh & McLennan.

The third party involved is the *underwriter,* who shares in the risk with the owner/user. These are the people who pay the owner/user if there is a failure. In turn, the owner/user pays premiums to the underwriters for their satellite coverages. Johnson & Higgins and International Technology Underwriters (INTEC) are two leading U.S. underwriters. Lloyd's of London is a well-known insurance underwriter in the United Kingdom. The owner/user, broker, and underwriter are the main participants in the space insurance industry, but they are not the only ones.

Other participants are the *bankers* and the *space agencies*. Bankers' loans for commercial space ventures could be profitable for them. One bank involved in such ventures is the Union Bank of Los Angeles. Major space

Figure 9.1
Participants in the Space Insurance Industry

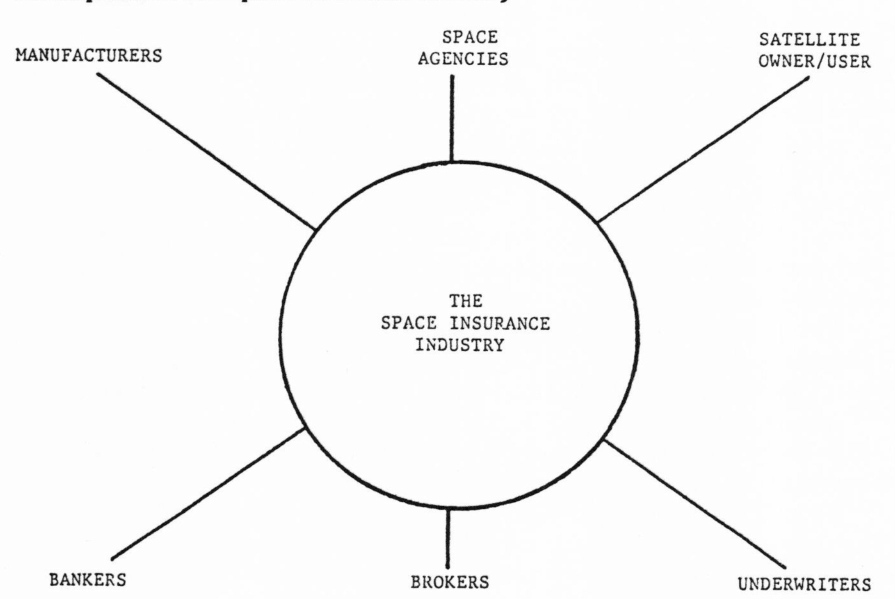

agencies are NASA, ESA, Japan's NASDA, the USSR's Glavkosmos, and the PRC's Ministry of Astronautics. One of the roles of space agencies is to place military satellites into orbit for national security; another is to process civilian payloads, such as launching commercial satellites or conducting scientific experiments in space. Table 9.1 lists major participants in the space insurance industry worldwide.

TYPES OF INSURANCE

Four broad but interrelated categories of space insurance are (1) satellite insurance, (2) insurance for space platforms, space stations, and other nonsatellite spacecraft, (3) insurance for equipment, products, and services on these nonsatellite spacecraft, and (4) life insurance for astronauts and workers in space. Other types of insurance will emerge with the passage of time, such as life and disability insurance for passengers on future (touristic) journeys into space; insurance for possible habitats in space, such as in lunar and Martian orbits or on the Moon; and insurance for lunar or Martian spaceports and future industrial complexes in space.

Except for satellite insurance and insurance for astronauts, these other types of space insurance are in their infancy, and some are just emerging.

In fact, one can argue that the entire field of space insurance is relatively new (beginning in 1965) and will continue to emerge and change. Since the subject of life insurance is self-explanatory and since insurance for nonsatellite spacecraft is in the embryonic stage, we look primarily at the most mature space insurance field, satellite insurance.

The satellite insurance coverages that prevail in the space industry are a product of over two decades of work. Space insurance for satellites generally consists of six coverages: prelaunch, launch, satellite life, third-party liability, liability transmission or revenues, and ground insurance. In most cases, the satellite owner is concerned only with prelaunch, launch, and satellite life. The user of the spacecraft or satellite is normally worried about transmission revenue.

The prelaunch part of a policy, or a separate policy, covers the client (normally the satellite manufacturer) on damage to the satellite during the manufacturing, storage, transportation, and launch-site assembly phases. It is usually called an all-risk physical damage coverage. The actual coverage terminates with the ignition of the launch vehicle. Also available is separate coverage for loss of revenue caused by launch delay or postponement.

Launch coverage is purchased for the beginning of ignition and continues until the spacecraft has reached its proper orbit. It usually lasts for as long as 180 days. This is the highest risk involved in the coverage and is where the majority of the premium is paid. The policy usually states that if there is a loss, the following will be paid: cost of a replacement spacecraft, cost of relaunch services, and lost revenues and extra expenses due to launch failure. Launch delay coverage indemnifies the satellite owner for launch delay costs, such as additional launch services, ground support, and lost revenues.

In-orbit satellite life coverage provides protection against partial or total satellite failure while in orbit. It goes into effect after the launch phase and usually covers a period of three years. Items that can be included in the coverage are protection from meteorite collisions and solar flares, and loss of revenues due to satellite or ground station failures.

Transmission insurance (in-orbit satellite life coverage) normally part of the in-orbit satellite life coverage, is included in the owner's satellite life coverage, but the lessee or user buys it separately. It is sometimes called user insurance. It protects the user from revenues lost due to failure of the transponders of the satellite and/or output power reductions.

Third-party liability coverage extends to all third parties for bodily injury and/or property damage during the launch phase and throughout the life of the satellite. An example of this type of coverage is that NASA insures the U.S. space shuttles for $500 million for third-party liability and requires that all payloads carried on board the shuttle carry third-party liability insurance.

Table 9.1
Major Participants in the Space Insurance Industry

SATELLITE OWNERS/USERS:

1. American Telephone & Telegraph Co. (AT&T)
2. AUSSAT (Australia)
3. Brazilsat (Brazil State Telecommunications)
4. Chinese Broadcasting Satellite Corporation
5. Arab Satellite Communications Organization (ARABSAT)
6. COMSAT (Communications Satellite Corporation)
7. Hughes Aircraft Corporation
8. Indonesian Government
9. Marslos (Mexican Telecommunications)
10. MCI Telecommunication Corporation
11. Telesat (Canada)
12. U. S. Military
13. Western Union & Telegraph Company

U.S. SATELLITE MANUFACTURERS:

1. Hughes Corporation
2. McDonnell Douglas Corporation
3. Minnesota Mining & Manufacturing (3M) Corporation
4. Radio Corporation of America (RCA)

U.S. BANKERS:

1. Morgan McPherson
2. Union Bank of Los Angeles

BROKERS:

1. Barring Aviation LTD (London)
2. Corroon & Black/Inspace, LTD (Washington, D.C.)
3. Crowley Warren, Co., LTD (London)

Table 9.1 (Continued)

4. Faugere & Jutheau (Paris)
5. Frank B. Hall, Inc. (USA)
6. Marsh & McLennan, Inc. (Hughes Corporation)

UNDERWRITERS:

U.S.

1. Cigna
2. International Technology Underwriters (INTEC) - Washington, D.C.
3. Johnson & Higgins (New York)
4. U.S. Aviation Insurance Group (New York)

Britain

1. Ariel Syndicate (Lloyd's of London)
2. Alexander Hiroden Underwriting LTD (London)
3. Merrett Syndicated, LTD (London)
4. Willis Fabes P.L.C. (London)

Indonesia

Asurani Jasa Indonesia (Jakarta)

SPACE AGENCIES:

Carriers

1. Arianespace (Paris)
2. National Aeronautics & Space Administration (NASA)
3. Japan's National Space Development Agency (NASDA)
4. European Space Agency (ESA)
5. Ministry of Astronautics (China)
6. Soviet Space Research Institute

Ground insurance coverage insures against property damage and business interruption on transmitting and receiving earth stations, transmitting towers, and headquarters. It is where the lowest risk is incurred.

Other miscellaneous insurance coverages associated with the operation of a satellite communications system are designed to meet a client's specific requirements, such as political risk coverage and video event coverage. The latter reimburses the insured for loss of revenue and out-of-pocket expenses when an event (such as a closed circuit sporting event) cannot take place.

The writing of space insurance is not much different from the writing of an average home owner or automobile insurance policy. Owners/users consult independent brokers, who negotiate the amount and types of coverages. Once the policy is agreed upon, the broker secures an underwriter to insure the satellite, although in reality, the broker gathers several underwriters because no one underwriter is willing to assume full risk.

Space insurance policies, however, are much more complex and lengthy than other types of insurance (home, boat, automobile, life). They are also more difficult and risky, for three reasons: (1) the potential claims are huge—generally anywhere from $20 million to $150 million; (2) predicting how many satellite launches will fail is impossible because of the varieties and complexities of satellites and launch vehicles and the many things that can go wrong with them; and (3) lack of expertise among underwriters, who have difficulty assessing and analyzing technological risk, thus potentially insuring very risky ventures at below-par premium rates.

This discussion has related to commercial satellites only, but there is a whole new class of payloads coming to the market for insurance. These are the materials processing payloads currently remaining within the U.S. space shuttle cargo bay but eventually to be deployed for subsequent recovery; and manufacturing and materials processing in space (MMPS) on space stations and space platforms. Clearly the risks are somewhat different than for satellites. Careful risk analysis and rating will be necessary if insurance support is to be obtained for these other ventures at affordable rates.

There will be a tremendous demand for space insurance if space commercialization proceeds. This is also a foreshadowing of the future need for space insurance by commercial participants (e.g., the United States, ESA, and Canada) on the U.S./International Space Station.

THE CURRENT SITUATION

Today the primary commercial space enterprise is communications satellites. Typically satellites are insured for three different phases of launch and operation: prelaunch, launch, and satellite life insurance. In addition, third-party liability insurance covers potential damage to persons or property on the ground in the event of a launch accident or failure. The first satellite insurance policy was written for the launch of International Satellite

Corporation's (INTELSAT's) Early Bird satellite in 1965 and involved prelaunch and third-party liability insurance. It was not until 1975, however, that complete insurance coverage for satellites was designed. That year, insurance was available for virtually any commerical communications satellite to cover loss of the satellite prior to reaching orbit, failure early in its operation lifetime, and third-party liability.

During the 1970s and early 1980s, there were many satellite losses (Table 9.2). During 1986–1988, there were many other satellite losses due largely to failures of *Ariane* (ESA), *Titan* (Martin Marietta), and *Delta* (McDonnell Douglas) rockets. These losses caused great concern among satellite insurance underwriters. In September 1977, ESA lost its *OTS I,* which was insured for $29 million. Perhaps the best-known loss was that of an RCA *SATCOM 3* satellite in 1979 when the upper stage apparently exploded, and a claim of $77 million was filed. By 1980, the loss ratio for insurance had exceeded 200 percent, causing a reevaluation within the industry. Many space insurance underwriters began to consider increasing insurance premiums from around 6 percent of insured value during the 1970s to as much as 20 to 30 percent. This did not occur, however.

A period of relative stability followed in 1980 and 1981. The years 1982–1985, however, were disastrous for insurance underwriters for satellites; during those years, several satellites were lost or rendered useless by failure to achieve proper orbit because of satellite rocket booster failure or malfunction of the satellite itself. For instance, in April 1982, India's *INSAT 1-A* became inoperable several months after it had achieved orbit. The resulting claim of $65 million was soon followed by the loss of the *MARECS-B* maritime communications satellite for ESA during an unsuccessful *Ariane* test launch, resulting in a $20 million insurance claim. Underwriters expressed concern that more disclosures were needed in the satellite and launch vehicle business. Insurance rates increased from 6 percent to 10 to 15 percent of value.

The advent of the U.S. space shuttle in 1981 was expected to improve launch insurance rates. Since the shuttle is a human-tended vehicle, it was considered more reliable than rockets because of its built-in redundancy. The expendable *Delta* and *Atlas-Centaur* rockets (launch vehicles) were also considered fairly reliable, with Europe's new *Ariane* vehicle the most risky because of its lack of experience.

The years 1984 and 1985, however, were particularly bad. Record losses were registered in 1984 for two satellites: *WESTAR 6*, owned by Western Union & Telegraph Company, and *PALAPA B–2* owned by Indonesia. Although these two satellites were successfully deployed from the space shuttle *Challenger* in 1984, they failed to reach proper orbit because their engines' upper stages failed. (Although the shuttle itself can get a satellite safely into LEO, another engine firing, using an upper stage or integral propulsion, is usually required to send the satellite to its proper orbit.) It

Table 9.2
Main Satellite Losses, 1977–1985

Date	Insured	Satellite	Failure	Insurance Loss*
Sept 13, 1977	European Space Agency (ESA)	OTS-1	Delta Rocket Explosion	$29
Feb 6, 1979	Japan's National Space Development Agency (NASDA)	ECS-1	Japanese Launcher	14
Dec 7, 1979	RCA	SATCOM 3	Apogee Kick Motor. Wrong Orbit	77
Feb 22, 1980	NASDA	ECS (AYAME)	Apogee Kick Motor	17
April 1982	India	INSAT 1-A	Delta Rocket Launch. Wrong Orbit	20
Sept 9, 1982	ESA	MARECS-B	Ariane Third Stage	20
Sept 9, 1982	RCA	SATCOM 2	Satellite	9
Feb 4, 1984	Western Union	WESTAR 6	PAM-D (Challenger Shuttle)	106

Feb 6, 1984	Indonesia	Palapa B-2	PAM-D (Challenger Shuttle)	75
June 9, 1984	Intelsat	V-F9	Atlas Rocket	102
Mar 8, 1985	TELESAT Canada	ANIK D-2	Satellite	5
Apr 12, 1985	Hughes	Syncom IV-3	Spacecraft	85
Apr 29, 1985	Hughes	Syncom IV-4	Spacecraft	85
Aug 1985		ARABSAT		25
Sept 12, 1985	GTE	SPACENET IV-4	Ariane Third Stage	85
Sept 12, 1985	ESA	SPACENET IV-3	Ariane Third Stage	65
			TOTAL	$819

*In millions of U.S. dollars
Sources: Economist 1985; Coleman 1985b.

became clear that regardless of the reliability of the space shuttle or other launch vehicle, other factors such as the reliability of the satellite itself are just as important as the launch vehicle for a successful mission.

The *WESTAR 6* and *PALAPA B–2* losses were soon followed by the loss of the *INTELSAT V-F9* satellite on June 9, 1984, when its newly modified *Centaur* upper stage malfunctioned after launch on an *Atlas* rocket. All three satellites were left in unusable orbits. The *INTELSAT V-F9* satellite was left to decay naturally; the *WESTAR 6* and *PALAPA B–2* satellites were retrieved in November 1984 on a special rescue mission by U.S. space shuttle (*Discovery*) astronauts and returned to Earth for refurbishment (by Hughes Aircraft Company) and relaunch in 1987. This impressive technical feat had little effect on satellite insurers, however, who had paid part of the cost for the satellite retrieval effort. The underwriters, Lloyd's of London, took title to the *WESTAR* and *PALAPA* and paid out $40 million.

The satellite losses or failures in 1984 and 1985 resulted in more than $600 million in losses to insurance underwriters (*Aviation Week & Space Technology* 1985l). These losses precipitated an increase in insurance premiums from an average of 10 percent of insured value to 20 to 30 percent of insured value, stricter policy terms, withdrawal of some underwriters in the United States and Europe from the space insurance business, a refusal to insure more than one satellite per launch, and the formation of a self-insurance subsidiary by Arianespace (the commercial arm of ESA) to add $40 million to $70 million of insurance capacity to satellite launches by the company's launch vehicle, the *Ariane.*

The explosion of the U.S. shuttle *Challenger* on January 28, 1986, just 73 seconds after liftoff, focused even more attention on space insurance—this time for satellites as well as for astronauts and others working in space. It also brought into sharper focus the dangers of space ventures. The explosions in 1986 of *Delta, Atlas,* and *Ariane* rockets carrying commercial and military satellites spotlighted again the problems of space insurance, the adequacy of inspection procedures and quality control, safety of space transportation, and space hardware reliability. Underwriters' confidence in the space industry was seriously damaged.

IMPLICATIONS FOR THE FUTURE

The high cost of space insurance and the difficulty of obtaining it will impede private sector space commercialization. For example, MPS, launching commercial satellites, building factories in space, and finding customers to use or buy space platforms (or parts thereof) will be inhibited by high space insurance premiums and the difficulty of obtaining space insurance. Fairchild Industries, for instance, was unable to find a single customer for its $200 million *Leasecraft* space platform after several years of marketing effort, in part because Fairchild could not guarantee potential customers insurance coverage or find insurance. Fairchild halted its program in November 1985.

Third-party liability insurance will also become imperative in space commercialization over time. Additionally the importance of life insurance for astronauts and future workers in space will grow, especially in the aftermath of the explosion of the *Challenger,* which killed all seven crew members aboard. But if space insurance claims continue to outstrip revenues from insurance premiums, the space insurance industry could collapse. The whole area of space insurance requires evaluations of policy options that the U.S. federal government, other governments, and the private sector will have to address if space commercialization is to proceed at a normal pace.

POLICY OPTIONS

Since high insurance rates or the unavailability of certain types of insurance will be disincentives for space commercialization, national governments will have to play an important role in ensuring that insurance can be obtained. In the past, the U.S. government has played a role in providing insurance when the insurance industry would not do so, for instance, it provides flood insurance and social security.

There are other forms government involvement in space insurance could take. It could subsidize insurance rates or insure the more risky phases of placing a satellite in orbit (launch and in-orbit testing) while the other phases are left for the private insurance market. Perhaps a government would provide insurance only for new, innovative space industries and not for the mature satellite communications industry. The government could establish a separate space insurance authority or agency to deal with space insurance issues, in the same vein as the Federal Trade Commission and the Food and Drug Administration, that would help to set quality standards for the design and manufacture of satellites, inspect satellites prior to launch, provide reliable information on the technological risks posed by the various aspects of space commercialization, and act as a fact finder in determining cause of loss. The government could also indemnify private insurers above a prescribed limit of losses per year (*Aerospace America* 1986) or provide tax benefits for underwriters in space insurance (*Space Commerce Bulletin* 1985).

There are also private sector policy options. One possibility is for companies to self-insure or to form self-insurance pools, as Arianespace is doing (*Aviation Week & Space Technology* 1985k). Self-insurance, however, is a disadvantage, primarily for small companies without substantial assets. Another possibility is changing the manner in which insurance is provided, with different rates for different phases of the space operations. This is essentially what INTEC (International Technology Underwriters) has done. It will no longer insure satellites before or during launch but will provide coverage once the satellites are in their proper orbits (*Aviation Week & Space Technology* 1985l).

A policy option that poorer nations may pursue in trying to defray the high cost of space and launch insurance is countertrade or barter. Countertrade normally involves a buyer who has some type of raw material or agricultural products, which are exchanged through a broker for the needed goods or services—in this case, space insurance. In some cases, the launch system operator may be willing to exchange the cost of the insurance premium for transfer of technology embodied in the satellite to be launched. Others could take long-term countertrade contracts for launch and insurance costs.

All of these policy options will be strengthened by increased ground testing of satellites by manufacturers prior to launch, improved launch vehicles, and "space warranties" by satellite manufacturers. Warranties would force satellite manufacturers to pay closer attention to quality control and the design and manufacture of satellites since they would be responsible for satellite replacement, loss of revenue to users, and so on as a result of faulty satellites.

These public sector and private sector policy options are starting points. They will require a great deal of work, debugging and improvement, and the collaboration of previously mentioned participants in the space (insurance) industry in order to come to successful fruition. These policy options will take several years of trial and error before they achieve maturity and will still continue to evolve as space commercialization unfolds.

SUMMARY

A key to the commerical development of outer spce is the availability of insurance coverage—not just engineering feats or marketing savvy. This chapter shows the role and place of space insurance in space commercialization.

Space insurance is a key to space commercialization by the private sector. Private corporations will not invest readily in space ventures or expose their expensive space assets to tremendous risks without insurance coverage. Furthermore, although today's capital markets have the capacity to finance large space systems, the willingness of the debt markets to supply funds for space ventures will be based heavily on the presence of insurance to protect against losses from launch failure and premature end of life.

The unexpected loss of many communications satellites, especially between 1984 and 1986, is having (and will continue to have) a profound effect on the space insurance industry. Premiums have escalated enormously, and many underwriters have withdrawn from the space insurance market. Future unavailability of insurance could endanger future space commercialization projects and retard space commercialization. It would also encourage more self-insurance, as done by Arianespace.

Future unavailability of space insurance or very high premiums will also encourage new risk takers—corporations willing to send up commercial satellites without insurance. RCA did this in November 1985 when it launched its $100 million satellite aboard the U.S. space shuttle *Atlantis*. RCA was unwilling to pay an insurance premium of $30 million (30 percent of the insured value of the satellite) that underwriters wanted. RCA won its gamble.

If private insurance companies cannot make a profit from insurance premiums, they will get out of the space insurance business. Whether or not they do so, national governments will be forced to intervene in order to control space traffic, implement space laws, and perhaps act as a space insurer of last resort. Lawyers will also become progressively involved in settling claims among owners/users, manufacturers, and underwriters.

As new fields of space commerce develop over the next decades there will be a growing need for new types of insurance. Around the year 2000 and beyond, demand for insurance will come primarily from seven different customer groups:

1. Communications satellite owners
2. Remote-sensing system owners
3. Free-flyer operators
4. Human-tended facility operators
5. Satellite servicing firms
6. MPS manufacturers
7. Commercial launch vehicle operators

Although insurance policies can be written to cover almost any contingency that would befall a commerical space firm, only a few types of insurance can be expected to generate reasonable or significant profits. These include the provision of insurance to cover (1) orbital maneuvering vehicles (OMVs), free flyers, and human-tended facilities (sometimes referred to as hardware insurance); (2) the value of materials produced in orbit (loss of revenue insurance); and, perhaps, (3) privately owned launch vehicles. The last two will not emerge until after the year 2000 because of the length of time required to gear up for these areas of space commercialization, attendant technological hurdles, and slow economic demand.

The space insurance industry is still in its infancy, but technology is rapidly evolving and costs are skyrocketing. Although the industry has suffered heavy losses, it may continue to support future space commercialization. The space insurance industry will also continue to face broad, challenging questions. If a U.S. satellite collides with a Soviet satellite, who pays, and how is compensation determined? Who pays when an insured satellite collides with an uninsured satellite? Will future workers in space find ready insurance on Earth? These, and other questions like them, will have to be answered as the commercialization of outer space evolves. Space insurance is essential to space commerce.

CHAPTER 10

Space Law

As space commercialization grows, so too will a relatively new body of law: space law. But the growth of space commercialization may be limited by restrictive space laws or the absence of appropriate space laws to encourage privatization of outer space. The purpose of this chapter is twofold: to discuss the major space laws and to examine how these laws will influence private sector participation in space commercialization.

MAJOR SPACE LAWS

There are ten major space laws, treaties, and agreements.

1. Treaty Banning Weapons Tests in the Atmosphere, in Outer Space, and Under Water (1963)
2. Outer Space Treaty of 1967 ("Magna Carta for Space")
3. Agreement on the Rescue of Astronauts, the Return of Astronauts, and the Return of Objects Launched into Outer Space, 1968 ("Rescue and Return Treaty")
4. Accident Measures Agreement, 1971
5. Prevention of Nuclear War Agreement, 1973
6. Treaty on the Limitation of Anti-Ballistic Missile Systems, 1972
7. Convention on International Liability for Damage Caused by Space Objects, 1972 ("Liability Treaty")

8. Convention on Registration of Objects Launched into Outer Space, 1975 ("Registration Treaty")
9. Agreement Governing the Activities of States on the Moon and Other Celestial Bodies, 1979 ("Moon Treaty")
10. Commerical Space Launch Act of 1984

Treaty Banning Weapons Tests in the Atmosphere, in Outer Space, and Under Water (1963)

This treaty, the first to regulate space activities explicitly, prohibits the peacetime detonation of nuclear weapons in outer space and the upper atmosphere. The treaty does not prohibit nuclear testing in wartime. Also, several nuclear states with space capabilities are not parties to the treaty.

Outer Space Treaty (1967)

This is by far the most important of the current space treaties, characterized by some as the Magna Carta for space. The most salient provisions of the treaty are that (White 1984, p. 47):

- International law and the Charter of the United Nations shall apply to space activities.
- Outer space and celestial bodies are the province of humanity and shall be used only for peaceful purposes and for the benefit of all people.
- Nuclear weapons, other weapons of mass destruction, military bases, and military maneuvers are banned from space.
- Outer space shall be free for exploration, use and scientific investigation.
- There can be no claims of sovereignty or territory by nations over locations in space "by means of use or occupation, or by any other means."
- Jurisdiction over space objects launched from Earth shall be retained by the launching state.
- Private interests are recognized as having freedom of action in space so long as a government or group of governments on Earth authorizes and exercises continuing supervision over their activities. Signatory nations (seventy-eight at last count, including the United States and the Soviet Union) are under a duty to oversee the activities of their citizens and commercial ventures in space.
- Governments are liable for damage caused on Earth by their space objects.
- Astronauts are "Envoys of Mankind" and are entitled to noninterference and all necessary assistance in distress.
- The natural environments of celestial bodies should not be seriously disrupted, and Earth must not be contaminated by extraterrestrial organisms.

Based on the Outer Space Treaty, does private industry have a right to participate in outer space activities? Generally yes. Early in the negotiations of the treaty, however, some nations recommended that participation in outer space activities be limited to nations. Private industry, it was suggested, should not be permitted to participate. Those suggestions were not adopted, and the treaty does not restrict participation to governments. In fact, Articles VI and IX of the treaty make specific references to "nongovernmental entities" and to the activities of a state "or its nationals in outer space." Article VI of the treaty, however, does require that "activities of non-governmental entities in outer space, including the Moon and the other celestial bodies, shall require authorization and continuing supervision by the appropriate State Party to the Treaty."

Agreement on the Rescue of Astronauts, the Return of Astronauts, and the Return of Objects Launched into Outer Space, 1968

This agreement, often called the Rescue and Return Treaty, requires all nations to provide assistance to a damaged spacecraft and its personnel and to return a downed spacecraft and its personnel.

Accident Measures Agreement (1971) and Prevention of Nuclear War Agreement (1973)

These two treaties oblige the United States and the USSR to refrain from interference with the attack early-warning systems of the other side, including space satellites (Reynolds and Merges 1985, p. 134). Some question whether both superpowers would adhere to these two agreements in the event of war and question also the stability-enhancing nature of attack early-warning systems. These two agreements, together with the important role that these systems played in containing the 1962 Cuban missile crisis, argue strongly for excluding early-warning and surveillance satellites from the arena of conflict in future crises.

Treaty on the Limitation of Anti-Ballistic Missile Systems (1972)

This agreement, between the United States and the Soviet Union, prohibits the development, testing and deployment of antimissile weapons systems on Earth and in space. Many people believe that both superpowers are violating this agreement by secretly developing and testing such systems. It has also been suggested that the SDI program violates this treaty.

Convention on International Liability for Damage Caused by Space Objects, 1972

This so-called Liability Treaty provides for redress of damages caused by a space object but does not clearly provide such redress for damages caused by persons to the property or employees of another in outer space. Private industry would have to rely on the ability of nations to consult and resolve the issues on the application of future, evolving space law and/or international law.

Convention on Registration of Objects Launched into Outer Space, 1975

The Registration Treaty, which emanated from the Convention on Registration of Objects Launched into Outer Space, requires all space objects to be registered to a nation. Thus the treaty requires any private industry space objects to be registered to a nation. It may be difficult or even impossible to determine the nation responsible for a specific abandoned man-made space object, or portion thereof, after many years in orbit.

Agreement Governing the Activities of States on the Moon and Other Celestial Bodies, 1979

The Moon Treaty came into being in 1979; so far, only fourteen countries have signed it[1] (Mereson 1984), among them Chile, the Philippines, Uruguay, the Netherlands, and Austria (Williams 1987, p. 146). The Moon Treaty is very controversial, as evidenced by these provisions:

1. A clear prohibition of private ownership of extraterrestial real estate, or resources "in place," and a designation of extraterrestrial resources as the common heritage of mankind (CHM).
2. The eventual establishment of an outer space regime (equivalent to an "international regime or space government") whose authority would be actionable and whose purpose would be to oversee and regulate the "orderly development and exploitation" of extraterrestrial resources.
3. A ban on all weapons (not just nuclear or mass destruction weapons) from celestial bodies, although this provision is not applied to Earth orbit.

The Moon Treaty and the Outer Space Treaty are similar in at least one aspect: both forbid the use of the Moon and other celestial bodies for military purposes. The Moon Treaty differs, however, from the Outer Space Treaty in two ways. First, the Moon Treaty's CHM provisions—similar to those of the controversial Law of the Sea Treaty—have aroused fears that its ratification by the United States would severely limit, if not prohibit, private commercial space ventures. This may be a destabilizing force for private industry participation in outer space activities. Second, the Moon Treaty calls for an international regime to regulate the commercial development of outer space. Thus, the right to exploit the natural resources of the Moon and other celestial bodies is not necessarily secured for private enterprise. Such rights would be governed by an international regime, the composition of which is indeterminable. Furthermore, to the extent that such a legal regime can be divined or determined, the security of private industry participation in space activities would be considerably reduced. Its right to participate, to retain its technology, or to retain and control the benefit derived from its activities would be subject to the economic, political, and nationalistic considerations of an international body or organization and its member states.

Private interests in the United States have other specific fears about the Moon Treaty. They worry that an outer space regime (international regime or space government) will tend more toward a one nation, one vote structure than toward the contribution-oriented organization of the World Bank or International Monetary Fund. They are concerned that the majority of countries might insist, as they have in the Seabed Treaty negotiations, that this proposed space administration not simply issue licenses without discrimination (perhaps for a nominal fee or small net profit percentage) but also deny or control uses of outer space, levy stiff taxes, and/or oversee equipment use and retrieval in free space.

Although the mentioned treaties and agreements are in force, their provisions are not strictly enforceable; moreover, the prospect of global government does not seem to lie in the near future. Given the different political ideologies in the world today and the difference in the level of economic development of many countries, a global government may never materialize. Additionally, space may be the common heritage of all humanity in a philosophical sense, but the resources actually developed by private industry should belong to the entrepreneur who had the vision and took the risk to obtain the benefit. From a practical point of view, the country that has the technology to explore and exploit the resources of outer space and the power to keep those resources will gain much of the real estate or "presence" in outer space.

Commercial Space Launch Act, 1984

The Commercial Space Launch Act was enacted by the U.S. Congress in 1984 to codify the government's policy toward space transportation regulation and management. The main purpose of the Act is the regulation and management of nongovernmental space transportation to ensure public safety, compliance with international obligations, and national security and to pursue foreign policy objectives. One of its implicit goals is to encourage the private sector development of commercial launch operations. The Act is administered through the Office of Commercial Space Transportation (OCST) in the U.S. Department of Transportation (DOT).

The Act has the following major provisions (Straubel 1987):

- Section 2605 prohibits the launch of a launch vehicle or operation of a launch site by any person within the United States or any U.S. citizen outside the United States unless authorized by a license issued under the Act. Launch activities of the U.S. government are exempt from the Act.
- Section 2606 brings launch sites under the DOT's control.
- The OCST is responsible for the licensing of all aspects of a private launch site except radio communications. The Federal Communications Commission (FCC) retains authority over use of the radio spectrum.
- Section 2616 gives the OCST the authority to investigate activities covered by the Act's provisions.
- Section 2617 makes violation of the Act unlawful, and section 2618 imposes civil penalities for such violations.
- Once a formal licensing application is filed, the OCST must make a determination within 180 days.
- Before the OCST will issue a license for an unmanned launch, the applicant must obtain mission approval and safety approval from the OCST. The mission review approval process is intended to protect the public health and safety, property, national security, and foreign policy interest. The applicant must supply information on launch plan, payload operation, and financial responsibility. Launch plan information includes a description of the launch vehicle, a flight plan, and list of unique hazards that might be posed by the launch or reentry. Payload operation information includes the nature and ownership of the payload. The required financial responsibility information includes either evidence of commercial third-party liability coverage or evidence of the purchase of a commercial surety bond.

  In the safety review process, the OCST determines whether the applicant can safely launch the proposed payload. It examines site location safety, operating procedures and their adequacy, personnel qualifications, and equipment adequacy. An applicant may go through the safety review process before or after the mission review process. (See Figure 10.1.)
- The OCST will consult with the Department of Defense and the Department of State on a variety of launch issues to ensure U.S. compliance with international agreements and spot potential foreign policy complications.
- The OCST is authorized to establish rules and requirements for the issuance of a launch license.

The Commercial Space Launch Act is a good start for the development of a viable private launch industry in the United States; it creates the framework for developing a workable regulating bureaucracy. However, in order to achieve the objective of developing a viable private launch industry, the

Figure 10.1
Department of Transportation and Office of Commercial Space Transportation's Commercial Space Launch Licensing Process

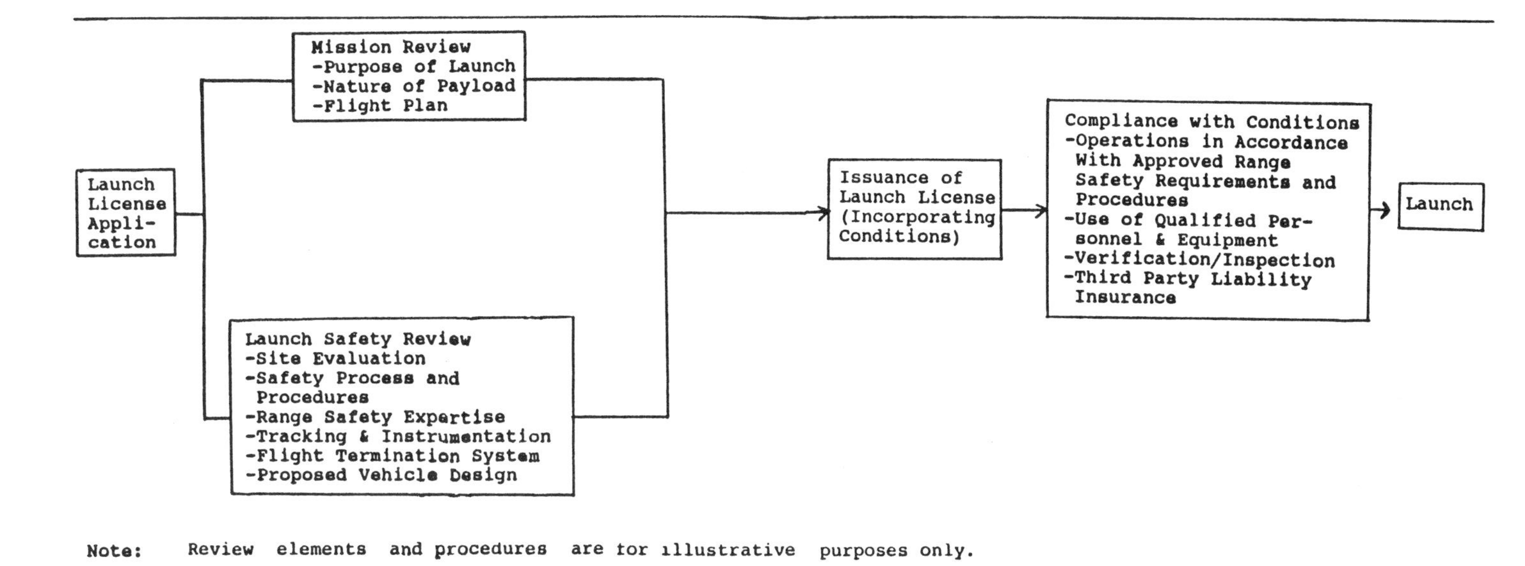

Note: Review elements and procedures are for illustrative purposes only.

Source: Straubel 1987, p. 956.

OCST should take specific steps in the short run. First, the OCST should have some of its own personnel capable of dealing with the technology of space transportation rather than relying exclusively on consultation with NASA or other agencies. Second, it should promulgate some minimum technical safety standards for the private launch industry to follow. The third short-run step is the creation of an environment conducive to the development of a private space transportation industry. For example, the federal government and NASA should announce a firm policy on NASA's future commercial launch activity, and, to create some certainty in the marketplace, a policy governing access to government launch facilities and charges for facility use should be enacted. NASA and the federal government are working out plans and procedures for private sector lease and use of government launch facilities.

SPACE LAW AND COMMERCIAL APPLICATIONS

These major space laws and treaties affect various aspects of space commerce.

Communications

Communications is the lifeblood of space activities. Without communications, there is no guidance, tracking, scientific data reception, or contact with astronauts. Three communications issues relative to space law are worth noting: the radio frequency spectrum, political issues of communications, and the geostationary orbit.

The radio frequency spectrum used for all communications is a finite resource managed by the International Telecommunications Union (ITU). UN resolution 1721 requires that "all states shall ensure that space telecommunications comply with ITU regulations" (Jenks 1965, p. 251). The ITU maintains a table of frequency allocations for satellite identification, telemetry, space research, and meteorological and radio navigation signals, among others. In addition, it establishes regulations that prescribe the power limits, angle of elevation, and power flux density limits to protect other services from interference (Jenks 1965, p. 253). This type of frequency management is essential when there are many communications satellites at varying orbit inclinations and altitudes.

In addition to the technical aspect of communications, space law addresses certain political issues of communications. For example, the 1967 Outer Space Treaty condemns propaganda designed to provoke or encourage any threat to peace, breach of the peace, or acts of aggression. With regard to the geostationary orbit, eight equatorial nations in 1976 asserted that their sovereign territories extend to the geostationary altitude. At this altitude, a satellite placed over the equator will match the Earth's rotation, thereby appear stationary, and will remain in constant view from nearly an entire hemisphere. These equatorial nations claim a violation of their air space, sovereign territory, and "right of privacy" from communications satellites in geostationary orbit (Christol 1984, p. 409). The United States disagrees with this viewpoint, opposing rigid planning of the geostationary orbit (Lowndes 1985).

Remote Sensing

Many countries, especially less developed countries (LDCs), have expressed concern over clandestine or secret remote sensing of their respective countries and the dissemination of such data (e.g., natural resources and military). They hold that the consent of the remotely sensed country should be required before data are collected or released. Space law does not yet address these issues directly. It is conceivable in the future, however, that the United Nations will adopt a data control procedure. As remote sensing moves into the commercialization era, Third World countries will question the issue of selling information about them acquired in "free space."

Ground-based Support

Most of today's space laws are really Earth laws designed to govern outer space activities. The majority of the activities involved in designing and adjudicating space laws are in the category of ground-based support. The fulfillment of contractual obligations is probably the most often addressed issue in space law. Space contracts—that is, contracts between aerospace companies involved in space activities—are governed by Earth law since those contracts are concluded on Earth even though the fulfillment of a contract may be in outer space. Examples are the delivery of one company's satellite to geosynchronous orbit by another company's booster and the relaying of one company's signal through another company's satellite.

Of all the ground-based support activities, the one law-related issue receiving most attention is that of space insurance, primarily because many insurance underwriters have lost millions of dollars from insuring satellites that failed to deploy or malfunctioned shortly after deployment. In one instance when a satellite failed to deploy, the insurance underwriter refused to honor its contract to cover the loss. The ensuing lawsuit by Hughes Aircraft sought to recover not only the cost of the satellite ($84.7 million) but also $25 million in punitive damages. The refusal of the space insurance company to pay was based on semantical interpretation of the policy coverage.

Industrial Research

The issues that relate to industrial research are primarily proprietary rights and the transfer of technology. These issues are not yet resolved and are at present to the advantage of NASA; that is; generally NASA has the right to technology and trade secrets that are part of the work it contracts out to be done. Recent legislation threatens the proprietary rights of products or processes developed at private expense for sale to the U.S. government. The threat is primarily to subcontractors that design products for prime contractors. The new laws—DOD Authorization Act of 1985, Small Business and Federal Procurement Competition Act of 1984, and Deficit Reduction Act of 1984—require all solicitations for bids or proposals to include offers of unlimited rights to all supplier data. Many aerospace and non-aerospace firms that do business with the U.S. government are trying to change these laws. This new legislation threatens small aerospace companies that have survived by capitalizing on trade secrets developed at private expense.

Materials Processing in Space

There are no specific provisions in space law relating to MPS other than the provision by the Outer Space Treaty of 1967 and the corresponding provisions of the 1979 Moon Treaty about the peaceful use of outer space. Any use of outer space must also avoid harmful contamination of the Earth's environment and interference with the activities of other states. In the future, however, as MPS grows, new space laws will have to be fashioned to deal with such issues as patent rights for products made in space, product liability for space-made products, worker's compensation for space accidents relative to MPS, tort and criminal law in space, FDA inspection of factories in space, and FTC regulation of the marketing of space-made products, such as advertising.

Space Transportation

The major national and international regulations governing or related to space transportation are the 1967 Outer Space Treaty, the 1968 Rescue and Return Treaty, the 1979 Moon Treaty, and the Commerical Space Launch Act of 1984. In the United States, the bulk of space transportation regulations come from the U.S. Department of Transportation, whose Office of Commercial Space Transportation (OCST) has the responsibility for clearing the path for private sector space transportation ventures. Among its chief concerns is safety. As private space transportation comes on line, new space laws concerning issues of space routes, liability insurance coverage, contracts between the government and the private sector, and others will have to implemented.

The Moon, Mars, Asteroids, and Beyond

One day space law may be a separate branch of law rather than a facet of international law. Such law would have to address issues such as mining the Moon and asteroids, building industrial and military bases on the Moon, tort law and criminal law on these and other celestial bodies, and on space stations. A few proposals relative to othese issues have been made.

One is for a lunar base, proposed by two scientists from the Los Alamos National Laboratory in New Mexico. The lunar base would include a spaceport(s) to handle incoming and departing spacecraft, refueling, and repairs; and other facilities, such as scientific laboratories, hospitals, and housing for astronauts, scientists, and other workers in space. The Moon base would provide its own energy sources and be a central base for varied Moon activities, such as mining and agronomy (Koltz 1983; David 1982). It could also act as an intermediary base for journeys to Mars and other distant planets.

The Moon base proposal would have the base's first module in place by the year 2000. This time frame, and the lead time required to develop the facilities, would put the base in competition with the U.S./International Space Station. Given the expense of space ventures and other obstacles, a more realistic time frame for a Moon base would be around the year 2010.

A Moon base would run into some political problems. For example, the Moon Treaty prohibits "national appropriation" of the Moon or any other celestial body. It is not clear if mining the Moon by national and private industry is allowable. Also, given that the Moon is the "common heritage of mankind," much like the seabed under international waters, some sort of allocation procedure would have to be established that would allot portions of the Moon's surface to those interests wishing to mine it. Such a procedure would be difficult to enact and enforce.

It is easy to see the importance of clarifying such questions as proprietary rights and technology transfer. The technical capabilities are bogged down in the UN by the LDCs, which fear that most of the benefits of outer space will be parceled out before they have the capability to exploit them. Their fears are justified; the spoils of outer space will go to nations with the technological expertise, economic strength, and political will to exploit them first.

The spoils of outer space will not be fully reaped until space stations are in place in outer space. Today the most prominent space station being designed and built is the U.S./International Space Station. Several legal challenges and issues need to be resolved before the station becomes fully operational in the first decade of the twenty-first century. Even then, new legal issues will evolve to challenge the best legal minds.

SPACE STATIONS AND THE LAW: SELECTED LEGAL ISSUES

The United States, ESA, Canada, and Japan are collaborating on the design and development of the U.S./International Space Station. This project will face many legal issues.

Jurisdiction over Space Station Activities

One of the fundamental legal issues facing space station partners concerns jurisdiction over space station activities. (Jurisdiction is a legal concept used to describe a state's (country's) right to take action, such as to prescribe and enforce rules of law, with respect to a particular person, thing or event.) The fundamental question to be determined is which country's laws will be used in governing activities on the U.S./International Space Station.

Station partners have at least four alternatives to choose from in setting up a legal system to govern the station (*Commercial Space* 1987c, p. 45). First, they could agree to allow one country legal jurisdication over and control of the station. Although this seems theoretically desirable, it will not be workable due to political considerations. If the United States, for instance, were to assert sole jurisdication over the station, other countries would withdraw their participation.

A second alternative, the extreme of the first, is to place each module of the station under the jurisdiction of the sponsoring country. This is not considered acceptable by U.S. officials, who want to avoid establishing national enclaves on board. Such a system is also unworkable in practice. National enclaves would be counterproductive to a basic principle of co-operation in the development and use of the station: equal access to all parts of the station by all partners.

A third alternative in jurisdiction over the space station activities is to place the entire station under the joint jurisdiction and control of several nations, in effect, making the station a multinational space station. Under this alternative, the United States, ESA, Japan, and Canada would jointly own and register the U.S./International Space Station through an international joint venture. Under current international law, joint registration (as distinguished from ownership) of space objects is not provided for (U.S. Congress 1986, p. 31).

Article VIII of the 1967 Outer Space Treaty establishes the principle that "a State . . . on whose registry an object launched into space is carried shall retain jurisdiction and control over such object." The 1975 Registration Treaty maintains that where two or more states may be considered "launching states," they shall jointly determine which one of them shall register the object, bearing in mind the provisions of Article VIII of the Outer Space Treaty. Under the Registration Treaty, then, participants in a joint space endeavor must choose which one shall be the registering state. Nonetheless, the Registration Treaty also states that such a joint determination is to be without prejudice "to appropriate agreements concluded . . . among the launching States on jurisdiction and control over the space object and over any personnel thereof." Therefore nations wishing jointly to own and exercise jurisdiction and control over a space station can follow the suggestion of the Registration Treaty to engage in an agreement separate from the actual registration.

A fourth option in establishing jurisdiction over the U.S./International Space Station would be to set up an international governmental organization, such as Intelsat, to have jurisdiction over the station. Assuming nations would wish to avoid some of the problems caused by concurrent national jurisdictions, it is possible that an international organization could be formed to own, operate, and register the space station. Since such an organization would not be able to develop a completely independent body of law to regulate space (station) activities, it would still be necessary to decide which national laws or combinations of national laws would apply to the space station organization. Such an organization could have quasi-legislative powers (subject to the concurrence of the member states) similar to those held by Intelsat. Such powers would allow the organization to make normal operational, management, and safety decisions without the need to renegotiate separate agreements among the member states.

Agreements used by the North Atlantic Treaty Organization (NATO) members to resolve questions of jurisdiction and control over troops stationed in NATO countries could be used as a model for the space station. The NATO Status of Force Agreements divide jurisdiction among member countries on the basis of whether the offense was civil or criminal, if it was committed on or off the military base, if it was committed while on official duty, and other factors.

Many individuals and organizations have different opinions about jurisdictional issues and potential legal problems relative to the operation of the U.S./International Space Station. The congressional Office of Technology Assessment (OTA) (U.S. Congress 1986) recommended in 1986 that Congress immediately start examining legal problems presented by multinational space station operations. The Science and Technology Committee of the U.S. House of Representatives has asked NASA to establish a panel of legal experts to review the U.S. position in station negotiations and comment on the possible negative effect legal provisions of the agreements may have on future use of the station by U.S. industry. Some members of Congress are concerned that U.S. interests will not be protected adequately in these areas.

Most of the legal experts who participated in a 1986 workshop on space stations and the law, from which the OTA report was published, were skeptical of the need for new international treaties or national space codes to cover station activities. They feel instead that problems should be addressed as they arise by applying international agreements, passing new congressional legislation, and using court decisions. Other legal experts who participated in the workshop disagreed. They think complex legal issues need to be resolved in advance, not dealt with after they have resulted in a case or controversy. It is recognized, however, that some legal issues will arise in management and operation of the U.S./International Space Station that will not be anticipated and will have to be dealt with on a case-by-case basis.

Both Congress and the judiciary will have some responsibility to sort out space station legal issues. Congress will be called on to decide which existing laws apply to space, what new laws are necessary to protect U.S. nationals living and working in space, and how best to encourage commercial activities on the space station. The courts also may set rules for space-related product liability, contract, intellectual property, and other issues that Congress has not decided.

Technology Transfer

The space station will set new precedents in the area of space law and the related field of technology transfer. For example, technology transfer laws will have to be refined due to the easy flow of data and other technology from one space station module to another. But space station partners would like to keep their data as proprietary as possible. Each partner has differences of opinion over contractual access and property rights aboard the station.

The Europeans want to construct a detachable or free-flier laboratory module, whose primary purpose would be MPS. The United States strongly objects to this demand, maintaining that such a module is not true participation by the Europeans in operation of the station. Instead, the United States believes it is a disguised attempt by the Europeans to learn about the specifications of the space station technology, which would allow them one day to separate the module from the station, fly away, and create their own station.

The Japanese do not object to a module permanently attached to the station; however, they disagree with NASA over proprietary rights of activities to take place on the station. They want at least 70 percent of all their work kept private (Anderson 1986). On the other hand, NASA's charter has always called for quick and early release of all space discoveries and experiments except in cases of national security. NASA feels that both the Japanese desire for privacy and NASA's open door charter can be dealt with in a manner satisfactory to all, although details of how NASA will resolve these conflicts have not been settled.

Complicating the Japanese privacy issue is the Canadian proposal. The

Canadians want to build a satellite servicing center aboard the station using technology they have developed in the space shuttle program. They then want to barter the service in exchange for information acquired in the research on materials processing. This is the same research that the Japanese want kept confidential (Anderson 1986).

Legal and technology transfer issues on the station are part of ongoing negotiations for agreements between the United States and its partners that will govern the program during construction and launch and for the first years of operation. U.S. agencies taking part in the talks are NASA, the State Department, the Department of Defense, and the Department of Commerce.

Patent Law Issues

Two basic patent law issues must be addressed and resolved if space commercialization is to proceed: how to protect the intellectual property rights of private sector firms and individuals working with the government in space and how to ensure that U.S. patent law protections apply to space activities.

Intellectual Property Rights in Government and Private Sector Space Activities. Section 305 of the 1958 National Aeronautics and Space Act (NAS Act) states that whenever any invention is made in the performance of any work under any NASA contract, such invention becomes the exclusive property of the United States unless NASA waives rights to it. Over the last thirty years, NASA has interpreted section 305 to apply only to activities that have as their main purpose the development of some new product or process for NASA. This interpretation leaves private firms in a quandary, however. They may wish to introduce a spin-off of their invention(s) into the marketplace but must first seek licensing approval from NASA unless other arrangements can be worked out. With respect to NASA–private sector joint ventures, it has been NASA's position that neither party assumes any obligation to perform inventive work for the other, and accordingly, each party retains the right to any invention that may be made in the course of the venture.

One way in which the U.S. government has sought to encourage private sector MMPS activities has been NASA's Joint Endeavor Agreements (JEAs). The intellectual property rights of the private participant of a JEA have, to date, been protected by the contract provisions of the individual JEAs. For example, in the first JEA, NASA and the McDonnell Douglas Corporation (MDAC) agreed that NASA would not acquire rights in inventions made by MDAC or its associates in the course of the joint endeavor unless MDAC failed to exploit the inventions or terminated the agreement or unless the NASA administrator determined that a national emergency existed involving a serious threat to public health.

U.S. Patent Law and Space Activities. There are two opposing views regarding U.S. patent law and space activities. One is that current U.S. patent law logically covers space activities. That is, activities aboard a U.S. spacecraft, for example, are tantamount to activities in the United States so inventions on a U.S. spacecraft should be protected by U.S. patent law. The other viewpoint, held by the U.S. Justice Department, for example, argues that it is not clear whether activities on a U.S. spacecraft could be viewed as activities in the territorial United States, and therefore, U.S. patent laws might not apply to such spacecraft. One thing is clear however: U.S. patent laws must be modified and extended to include activities in outer space. Otherwise private sector participation in space commercialization will be inhibited. To the extent that patent law problems could limit the success of the U.S./International Space Station, every effort must also be made to achieve some type of international coordination. Such coordination should seek to resolve issues of secrecy and which country's patent laws should be applied whenever non-U.S. nationals invent products or processes aboard a U.S. spacecraft.

Criminal Law in Space

Criminal law in space is another issue that must be addressed sooner or later with regard to the space station or general activities in space. Two interrelated issues are important: What body of criminal law is to be applied? and How are the relevant laws to be enforced?

The simple answer to the first question is, "whichever nation has jurisdiction and control over the space object." However, questions of jurisdiction are not easily resolved without first knowing how the space station is to be owned and registered. For example, in the case of the U.S./International Space Station, U.S. law could be relied upon, some type of shared jurisdiction and control scheme could be used, or more than one body of criminal law could be applied. In the light of potential difficulties, it might be desirable to negotiate an agreement in advance of occupying the space station.

The answer to the second question—How are the relevant laws to be enforced?—is that any such enforcement may be difficult and should be worked out ahead of time. Suppose that a British astronaut assaults a U.S. astronaut while the U.S. astronaut is in the British module. The United States has the jurisdiction to prohibit such conduct, but whether it would have the jurisdiction to enforce such rules would depend somewhat on whether it had some prior agreement with the British government. Lacking an agreement with the British government, the United States would not have jurisdiction to enforce these laws in the parts of the space station under British jurisdiction and control. Both governments would perhaps have to resolve the issue through diplomatic and other negotiated means.

Tort Law in Space

As people begin to live and work in space, incidents of damage caused by intentional actions or negligence are certain to occur. Individuals or organizations seeking compensation for damage to property or personal injury may look either to international space law or to the tort laws of their own or other nations. Neither of these courses of action is without difficulty. Current international space laws are little more than agreed fundamental principles, and no efficient mechanisms exist for applying these principles to specific cases. National tort laws, on the other hand, are well developed but vary from country to country.

International Law. The Outer Space Treaty provides that states bear international responsibility for national activities in outer space and that the activities of nongovernmental entities (individuals and corporations) shall require authorization and continuing supervision by the respective state. The treaty also declares that a launching state is internationally liable for damage to another state or its people. The 1972 Liability Treaty restates and expands on these principles and specifies procedures for making and settling claims.

Although the Outer Space Treaty and the Liability Treaty establish several key principles, such as international liability for damage on Earth or in space, both treaties leave many questions unanswered. Three important problems raised by the current international space liability regime are the uncertain applicability to activities aboard space stations; the lack of attention to damage caused by, and the liability of, individuals; and the absence of an efficient mechanism for resolving disputes between individuals.

Both the Outer Space Treaty and the Liability Treaty focus on damage caused by space objects rather than on damage caused by individuals in space. This is understandable because the primary concern of the drafters was probably to offer some protection from falling or colliding space objects. The crash of the radioactive Soviet satellite *Cosmos 954* in northwestern Canada was an example of the kind of injury to property best suited to the protections of the international treaties. On a space station, however, individual personal injury actions resulting from intentional actions or negligence are likely to occur. Such issues need to be addressed before the U.S./International Space Station becomes operational.

With respect to the absence of an efficient mechanism for resolving disputes between individuals, serious questions exist as to whether current international laws could be applied to assist individuals. The Outer Space and Liability treaties establish no cause of action, no courts, no rules of procedure, and no method of enforcing even agreed resolutions. Lacking such mechanisms, claimants are forced to rely on the diplomatic procedures commonly used between nations.

But Article VIII of the Liability Treaty requires that the state—not the injured person—present the claim to the launching state—not the person. Because nations and not individuals are involved, claims for compensation must be presented "through diplomatic channels." If the two states in question do not have diplomatic relations, the claimant may present its claim through another state or through the secretary-general of the United Nations. Assuming that a claim has been filed and diplomatic relations have failed for a year, then Article XIV of the Liability Treaty authorizes the parties to set up a claims commission composed of three members (the two parties and an agreed chairman).

National Tort Laws. With regard to space activities, tort laws present two basic obstacles: they generally do not apply to (extend to) space activities, and they vary from country to country. The development of space law may take care of these two obstacles by providing a body of space law for reference. Having acknowledged these two obstacles, it is necessary to inquire which domestic laws would be applicable to a given case.

Every nation has its own methods for choosing the law applicable in a specific case. The most common of these are the *lex loci delecti, lex fori,* and the law of the state having the greatest interest. *Lex loci delecti* refers to the law of the place where the offense occurred. Outer space, however, is not subject to national law and has no clear "law of the place." Whether the *lex loci delecti* rule can be applied to the space station will depend on how nations agree to exercise jurisdiction and control over the space station.

Lex fori refers to the law of the forum where the case is brought. This approach could be used on the space station but would depend on how questions of jurisdiction and control are resolved. Regarding the law of the state having the greatest interest, this rule—probably the prevailing U.S. standard—looks to which state's contacts with the incident are the most substantial and applies the relevant laws of that state. Because of its flexibility, this rule could have the greatest applicability to space station activities.

Given the current level of space activity, another solution to the problem of liability might be to negotiate interparty waivers of liability. The limitation of such agreements is that they cover only signatories. Interparty waivers of liability were used, for instance, in the 1973 Spacelab Agreement and are regularly used in shuttle launch agreements. Article 11(A) of the Spacelab Agreement, for example, provides that the United States "shall have full responsibility for damage to its nationals ... [resulting from] ... this agreement." ESA nations accept a similar responsibility under this article. In other words, the United States would not sue ESA for damage to U.S. nationals or property, and vice-versa. However, Article 11(C) of the Spacelab Agreement acknowledges that in the event injury is caused to persons not party to the agreement, "such damage shall be the responsibility of ... [the United States or ESA] ... depending on where the responsibility falls under applicable law."

FUTURE CONCERNS

The basic space laws discussed form a good foundation on which to build other space laws. For the private sector, however, two additional issues are of particular concern in space activities: product liability and export law.

Product Liability

As space research and commerce grow, so is the likelihood that people will eventually be injured or killed on the space station by products manufactured on Earth; on Earth by products manufactured on the space station; and on the space station by products manufactured on the space station. With the passage of time, product liability is destined to become an important issue. The Outer Space and Liability treaties discuss damage caused by space objects in a way that applies to states and intergovernmental organizations but has little relevance for private citizens. National product liability laws, on the other hand, apply to individuals but vary not only from country to country but within the regions of individual countries. For this reason, some legal experts feel that there is no clear legal recourse at present for individuals injured or killed on the space station.

In addition to conflicting national laws, the uncertain nature of space jurisdiction and the possibility of multiple jurisdictions make the choice-of-law question extremely difficult for space station product liability cases.

There are three multilateral instruments currently in force on product liability cases on Earth: the Hague Convention[2] to determine applicable "conflict of law" rules, the Council of Europe Convention,[3] and the European Economic Community (EEC) Directive.[4] These instruments offer a little guidance on how to resolve similar problems that might arise on the space station. For example, nations could, following the EEC Directive, enter into an agreement to modify their national laws to adopt a strict liability standard of proof for all product liability cases arising from the space station. In addition, such an agreement could allow nations to establish a ceiling on financial settlements. In the twenty-first century and beyond, an international treaty on product liability as it pertains to space stations may become reality.

Export Law

Export law is complex; export law as it may pertain to space becomes even more complex. For example:

- Depending on how jurisdiction is allocated on the space station, should transfers between national modules be regarded as imports or exports?
- Assume the shipment of made-in-space products to Earth. What would be the effect of the jurisdiction of the modules? the nationality of the producer? the fact that the product might first land in the United States on the shuttle and then be shipped to the ultimate destination?
- Will some countries bar the import of certain space-made products for fear of importing organisms peculiar to outer space and dangerous to humans?
- Under current plans, components ultimately destined for the space station will be manufactured in many countries. It is important to develop rules that allow the easy transfer of space station components between nations.

SUMMARY

There are at least ten major space laws, treaties, and agreements, and they will influence private sector participation in space commercialization. The most important of these space laws are the 1967 Outer Space Treaty, the 1968 Rescue and Return Treaty, the 1972 Liability Treaty, the 1975 Registration Treaty, the 1979 Moon Treaty, and the Commercial Space Launch Act of 1984. This chapter discusses the main provisions or tenets of these laws and their possible influence on space commercialization.

Generally space laws promulgate the following three principles:

1. Outer space and celestial bodies are the province of all humanity and shall be used only for peaceful purposes and for the benefit of humanity.
2. Governments are liable for damage caused on Earth by their space objects.
3. Private interests are recognized as having freedom of action in space so long as a government or group of governments on Earth authorize and exercise continuing supervision over their activities.

Space law will evolve like sea law: slowly and by trial and error. It will determine and influence private sector participation in the following areas of space commercialization: communications, remote sensing, ground-based support, industrial research, MPS, space transportation, and living and working on space stations. These areas of space commercialization will not be without difficulties and conflicts among firms, as well as among countries. For example, there are controversies regarding ITU's allocation of positions in the radio frequency spectrum, and many Third World countries question the secret remote sensing of their territories by industrialized nations such as the United States, regarding it as a violation of their free space and privacy. Firms involved in space commercialization are concerned about such issues as patent rights for products and processes developed in space, product liability for space-made products, and uncertainty about the vigor and alacrity with which the U.S. government will support space commercialization.

The development of the U.S./International Space Station will perhaps have the most profound effect on the development of space law. It will force world attention to many legal issues, such as jurisdiction over space station activities, technology transfer, patent law issues, criminal law in space, tort law in space, and product liability.

The commercialization of outer space is, by its very nature, a long-term process. So too is space law. Space law will take many centuries to develop. In the meantime, societies that are the strongest economically and that have the staying power and political will to explore and exploit outer space will garner most of the spoils of space commercialization and establish space laws in their own economic interests.

NOTES

1. The United States has signed but not ratified the treaty, although it helped draft it.
2. The United States is not a party to this treaty.
3. Ratified by only three members.
4. In force beginning in 1988.

Conclusion

Many books have been written about space. They range from personal accounts of journeys into space, such as Joseph Allen's *Entering Space: An Astronaut's Odyssey* (1984), to L. B. Taylor's *Commercialization of Space* (1987). Thousands of articles have also been written on the subject, appearing in such magazines and journals as the *Futurist, Space World, Science Digest,* the *Economist,* and *Business Week*. Some magazines, such as *Aviation Week & Space Technology,* specialize in the topic. Space business is big business and becoming bigger every day. This book has examined topics that space businesses or those thinking of entering that business will find informative and useful, among them space operators, materials processing in space, spin-offs of space technology and market opportunities they have created, how to identify customers for orbital facilities, how to market to NASA, space insurance, and space law.

Let us look now at some other interesting issues relative to space commerce.

SEARCH FOR EXTRATERRESTRIAL INTELLIGENCE

The search for extraterrestrial intelligence is not just the subject of science fiction books and movies; it is also part of space business. It involves, for example, designing, building, and marketing special satellites to collect information, sounds, and signals from space; larger and more powerful telescopes for astronomy studies; special-purpose computers powerful enough to collect, compare, analyze, and store millions of signals per second; and more earth stations for data collection from space, data production, and analysis.

In the United States, one of the leading organizations involved specifically in the business of scientific searching for extraterrestrial intelligence (SETI) is the Oak Ridge Observatory, Harvard, Massachusetts, which is run by the Smithsonian Astrophysical Observatory. Since 1984, the 84-foot radio telescope at the Oak Ridge Observatory has been scanning the heavens, searching for signals from intelligent life in the galaxy. Harvard University physics professor Paul Horowitz runs the project. NASA also operates two SETI programs: a targeted search at the Ames Research Center at Moffett Field, California, and an all-sky survey at the Jet Propulsion Laboratory at the California Institute of Technology in Pasadena (Dall'Acqua 1988, p. 18). Since the beginning of scientific SETI in 1960 by Frank Drake (then of the National Radio Astronomy Observatory, Green Bank, West Virginia, and now an astronomy professor at the University of California, Santa Cruz), about fifty large-scale SETI programs have been conducted, many in the Soviet Union.

THIRD WORLD ACTIVITIES IN SPACE?

When we think of space commercialization, the large industrial nations come to mind: the United States, Japan, West Germany. Third World countries by and large, are too poor to be involved seriously in space commerce. They are more concerned with solving pressing economic and social problems of poverty, high unemployment, illiteracy, and low-growth economies. They tend to be buyers of space technology rather than producers. As such, Third World countries, like India, Pakistan, Mexico, and Brazil, are markets for space goods (e.g., satellites) produced by the world's space commerce leaders. A $50 million or $100 million satellite can serve the communication needs for an entire country. The individual country can determine itself how to use these systems within its own culture. For example, one satellite in India beams messages and educational programs to 4,000 remote villages, as do similar satellites in Mexico and Pakistan.

The PRC and Israel appear to be two Third World countries that are exceptions to the rule; they are becoming actively involved in space commerce. The PRC is engaged in commercial space transportation and launch (Gengtao 1987) and beginning research into other facets of space commerce like MPS, space stations, and space platforms. It would not be surprising if China becomes a powerhouse in space technology and space commercialization within the next century. Israel is collaborating with the United States on SDI research and development, as are other countries, such as the United Kingdom.

THE SOVIET UNION AND THE UNITED STATES: COOPERATION IN SPACE?

Another interesting question is whether there will be meaningful cooperation between the Soviets and the United States in space ventures. That is doubtful because of sensitive technology, general mistrust of each other, and politics. The *Apollo-Soyuz* docking in space had some public relations value but did not represent a real collaborative space venture between the USSR and the United States. In May 1977, NASA and the Soviet Academy of Science signed an agreement for a shuttle-*Salyut* project, but in 1978, the Carter administration tabled it. Current NASA officials say there is no reason why the USSR and the United States cannot cooperate with each other in space ventures. Actually accomplishing it, however, is another story. Some scientists are putting forth the idea of a cooperative venture between the Soviets and the U.S. in a mission to Mars. There will not be meaningful collaborative space ventures between both superpowers during the next several decades, given their different political ideologies, mistrust, and efforts to protect their technological advancements.

THE NEW FRONTIER

The new frontier of space is sometimes compared to earlier frontiers of human settlement. Is this analogy too facile? Yes. There will not be a pioneering-type exodus into space comparable to the emigration of Europeans to the North American continent. Why? Satisfaction and familiarity with the Earth. Safety concerns. Health concerns about living in space for extended periods of time. The new frontier of North America also offered unlimited real estate, water, air, game, fish and forests—essential resources for life not found in space, at least not very easily. For quite a while, the early phase of space colonization will resemble scientific outposts in the Antarctic. The early space outposts may not even be permanently staffed but only intermittently manned. For example, a camp on the Moon might be occupied by thirty or so scientists for three months out of the year. However, as these outposts develop greater self-sufficiency, the length of time spent in space will increase. One day some may stay permanently.

THE NEXT FIFTY YEARS

The United States faces many critical economic issues: the federal budget deficit, exchange value of the dollar, unstable oil prices, trade barriers and imbalances, Third World debt, high levels of corporate and consumer debt, tax policies, and a propensity to give billions of dollars in aid to foreign countries while denying similar help to many domestic programs. How bright are the next fifty years for the American economy? Will twenty-first-century America be able to afford investment on the space frontier? Will it choose to do so? If it does, what are some of the possible accomplishments?

A number of problems will continue to plague the U.S. economy. Although they appear manageable in the light of the demonstrated long-term vitality and adaptability of the U.S. economic system, they will adversely affect the space program by diverting money and attention away from it. New presidential administrations will interrupt the smooth continuity of the space program and retard space commercialization. Other space accidents, like the *Challenger* tragedy, and ELV failures will set back the U.S. space program. Typically each accident like the *Challenger* sets back space commercialization about five years.

If we assume 3 percent annual real growth in the U.S. gross domestic product through 2040, modest population growth, major advances in science and technology, and growth in world trade, it is reasonable to project by 2040 a population of 300 million people, a gross domestic product of $14,000 billion, and a GDP per capita of $50,000 in today's dollars. With these growth figures, the industrial economy, and political will, the United States should be able to afford to lead the rest of the world on the space frontier. However, by 2040, Japan and the PRC may each be almost equal to the United States on the space frontier, poised to assume leadership in space if the United States falters. And all these countries will be vying with the USSR for supremacy in space.

There will be a number of major accomplishments in space:

- Full operation of the U.S./International Space Station and other space stations and space platforms built and owned individually by Japan, the PRC, ESA, and the USSR.
- Fleets of space transportation vehicles for journeys to LEO. The United States, the USSR, the PRC, Japan, and ESA will have such vehicles.
- Electronically sophisticated space probes that will map the resources of the Solar System and portions of deep space.
- Lighter and more powerful satellites of all kinds (direct broadcast, military, navigational, remote sensing, weather), each with a longer life span. Many such satellites have a life span of five to ten years before they malfunction and decay in space and have to be replaced.
- Spaceports in LEO and lunar and Martian orbits.
- Robotic factories in space for experimental MPS.
- Advances in human biological research (space medicine) aimed at facilitating humans living in the microgravity environment of space for extended periods of time. A corollary of this is that we need to settle the question of whether we should go to Mars in a zero-gravity vehicle or one with artificial gravity. The United States will probably have vehicle concepts for both.

Many experts in a wide variety of fields have projected other significant achievements during the next fifty years, which give clues to technology needs of the future and future markets for space commerce and nonspace commerce (*Futurist* 1988a):

- Commercially manufactured superconductive material at liquid nitrogen boiling point
- Space tourism
- Robot nurses caring for elderly and bedridden patients
- A retarded aging process
- An artificial eye
- Cures for many forms of cancer, for bone thinning in astronauts (with treatment applicable to people on Earth), for diabetes, and for AIDS
- Prediction of volcanic eruptions and earthquakes
- Voice-activated typewriters
- Aerospace planes

Two themes seem to predominate among these predictions: (1) engineering and technological advancements and (2) advances in medical science. It is instructive to take one of the predictions and see how cognizance of it is helpful to firms in the aerospace and nonaerospace industries in their preparation of future business plans. Let us look at the prediction for an aerospace plane.

The aerospace plane is an aircraft that will be capable of horizontal takeoff from a very large runway, travel at hypersonic speeds in the atmosphere, and return for a conventional airport landing. Travelers will be whisked in two hours or less from an airport runway on one continent to another halfway around the world. NASA and the U.S. Department of Defense are engaged in a joint project to develop such a plane (*Futurist* 1988c) for military purposes, with obvious spin-offs to the air passenger transportation market.

The plane will create a multitude of business opportunities and needs. They include research on and production of critical technologies, such as aerodynamics, propulsion, materials, and computers. One of the goals of the program is to produce by the year 2000 civil and military aerospace vehicles capable of traveling up to Mach 25 (twenty-five times the speed of sound) and commercial vehicles traveling up to Mach 12. Since today's commercial aircraft cannot withstand the heat and pressure of acceleration and flight at Mach 25, new structural material and designs are being developed that will be stronger and lighter and will find ready markets in commercial air transportation. (NASA and DOD can provide details regarding research needs and contract opportunities relative to the aerospace plane.)

One can go through similar scenarios for the other predictions. For example, research and development relative to robots will mean advances in, and new or expanding markets for, tele-operated or autonomous robots, expert systems, and artificial intelligence. Rockwell International, General Dynamics, and IBM are examples of some U.S. companies engaged in one or more of these areas. The introduction of voice-activated typewriters will result in one of the most lucrative consumer and industrial markets in the world.

THE MARKETING IMPERATIVE

Space business is long term, capital intensive, and risky. It requires monumental marketing efforts to be successful.

The Market

The first crucial step in any business, especially in space business, is to do a market assessment. Is there a market for our product? If so, how large is it in terms of appropriate units like number of firms, households, or individuals? Where is the market geographically located? What is the demographic profile of the market(s)? What is the likely demand for the product during each of the next five years? What are the potential revenue streams from such economic demand? Is the cost of getting into the business worth it?

Answers to these questions can be obtained by doing the following (not an exhaustive list and not in any order of importance):

1. Carry out the preliminary economic analysis internally before discussing the project with NASA or any other organization. This preliminary economic analysis will give the company a better idea about the feasibility and other aspects of its own project.
2. Contact the Office of Space Commercialization, NASA Headquarters, Washington, D.C., and/or other specific NASA locations for as much literature as possible on the type of project the company is pursuing. At the same time inquire about relevant NASA seminars, studies, and other publications, which can be excellent sources of information on the state of research in a particular field, market assessment, seed funding, and so on.
3. Contact the Superintendent of Documents, Washington, D.C., and inquire about other publications relevant to the project. Many NASA publications are sold at nominal prices through the superintendent. Other useful sources of secondary information can be found in public and university libraries. They include the

annual *Statistical Abstracts of the United States*, *Census of Business* and other useful censuses.

4. Contact one of the many NASA-related, university-based industrial applications centers that act as inexpensive consulting centers for space commercialization projects (table 5.2).
5. Contact a professor at a local university, and request that a market assessment study be done by his or her graduate students. The students benefit from exciting hands-on field and library research, and the company obtains low-cost, quality research.
6. Use other consultants, such as Booz, Allen & Hamilton or the Center for Space Policy, Cambridge, Massachusetts, for further studies.

Competition

In addition to a market assessment, the company has to make a competitor assessment. Who are the major competitors? What are their strengths and weaknesses? How long have they been in the particular business? The market assessment study should help to reveal major competitors, as will reading sources of information listed in Appendix A and perusing Appendix B.

Corporate Decision Makers

An essential part of a marketing program is knowing the key corporate decision makers who should be approached in the target company and to whom proposals should be addressed. If personal selling is required, careful analyses of the target company's corporate structure and its decision makers must be done prior to company visitation. This increases marketing efficiency. In many cases it may be necessary to approach more than one department or organization within the target company. For example, suppose the maker of a space platform (company XYZ) wants to lease space and utilities on its space platform to the 3M Corporation. Company XYZ will have to market its proposal to at least two groups of people at 3M: a marketing group and a science and technical group, which normally decide which projects to proceed with. Similar scenarios are the norm at other firms in space business, such as RCA, Du Pont, TRW, and Rockwell International.

Advertising and Promotion

Advertising is the engine of marketing. It is necessary to make the market aware of your company's space products and services, comparative advantages over competitors, and so on. Choosing the advertising media follows logically after choosing the target market. An effective marketing approach may be to advertise in different professional trade journals, some of which are mentioned in Appendix A. Choose the ones read by the target market. Draw up a media plan to cover such topics as types of media in which to advertise, ad content, ad size, frequency, and cost. Implement the media plan and evaluate its effectiveness. Take corrective action if necessary. An evaluative or monitoring mechanism is to include a toll-free telephone number or an address for additional information about the company's products or services.

In terms of promotion, the company should develop high-quality, informative literature, such as color brochures or small booklets. The quality of the literature reflects the company's reputation. Different brochures may be necessary for different products and markets.

Effective marketing also calls for professional (sales) personnel to make presentations to different professional organizations. These sales personnel should be knowledgeable about the product, articulate, and represent the company well. They should also be knowledgeable about the space environment and be able to answer a variety of questions on areas such as MPS, costs, potential markets, and potential profits.

Lobbying

In addition to the normally accepted marketing strategies, such as advertising and promotion, a space enterprise should also consider lobbying efforts directed toward the U.S. federal government, key personnel within NASA, and pertinent state government officials. Lobbying, an ongoing, long-term strategy, should emphasize, among other things, the benefits of the company's space venture to the government, NASA, and the public and the importance of space enterprise in general to the long-term technological leadership of the U.S.

Environmental Scanning and Analysis

In order to do an effective marketing job, companies in space business must continually monitor the economic, political, and legal environments. A booming economic environment is generally a stimulus to space commercialization; a sluggish economy is anathema. The state of the political environment will also determine the rate at which companies become involved in space business. For example, during the Reagan administration, space commercialization got a boost, especially with SDI projects. Will the next U.S. administration support space enterprise with the same zeal? Will it support some efforts and not others? For instance, the Reagan administration vigorously supported SDI projects but did not provide direct financial support necessary to encourage growth of the MPS industry. Lobbying the federal government to provide direct financial support for MPS activities appears appropriate.

The space enterprise should also monitor the legal environment in which it operates. For example, if it is in the MPS industry, it should monitor laws in that field and try to have input into new laws and regulations that will be written. As part of its scanning of the legal environment, the MPS firm should also review litigation involving firms that lease terrestrial manufacturing facilities and extrapolate from this potential litigation that may be encountered in the lease of an orbital manufacturing facility.

PRIVATIZATION

Privatization is expanding (Goodrich 1988). There is serious discussion at NASA headquarters in Washington, D.C., and among some U.S. federal government officials concerning moving toward privatizing space transportation. Currently space transportation vehicles, specifically the space shuttles, are owned by the U.S. government, which operates these vehicles and sells space transportation services to the private sector. Privatizing space transportation would mean that the government would get out of the space transportation business. It would allow the private sector to take full responsibility for financing and building space shuttles and other space transportation vehicles, and the government would become a purchaser of space transportation services rather than a seller.

Privatizing space transportation—making it basically similar to airline operations in the United States—is, however, easier said than done. The government would have to sell (or lease) its three remaining shuttles. But to whom? At a price tag of around $4 billion each (in 1988 dollars), no one corporation is in a financial position to buy one shuttle. Neither will a consortium of a few of the richest corporations in America. Which group of financial institutions will want to risk lending such a large sum of money on a risky, unproved business? None. The space insurance industry will not want to insure such vehicles, especially in the wake of the *Challenger* disaster. Most important, the market for space transportation is in its infancy; it will not mature until several decades into the twenty-first century. Since space business, like any other business, is concerned about the bottom line, companies will not want to buy or build space transportation vehicles for which there is little sustained economic demand. So for all these problems, such as cost and limited economic demand, privatizing the provision of space transportation will not materialize until thirty or more years from now. Space powers like the United States use space a great deal for military surveillance and defense purposes. It would seem prudent, therefore, for the government to have its own vehicles to launch its own sensitive defense satellites. Partial privatization seems more logical, therefore, than complete privatization of space transportation.

Space business is long term, risky, and fascinating. Someday it may become as routine and pervasive as the computer and aviation industries are today. Space commercialization is part of human evolution.

Appendix A

Space Business Publications

There are numerous publications on the space industry. This appendix provides a sampling of the important ones, listed alphabetically in order of their frequency, from daily to special events.

DAILY

Aerospace Daily, 1156 15th Street, N.W., Washington, D.C. 20005. Published by Business Publications division of the Ziff-Davis Publishing Co.

Newspapers, such as *The Los Angeles Times, The New York Times, The Wall Street Journal, The Washington Post, Houston Chronicle,* and *Miami Herald* occasionally run pieces of interest. Topics can be traced through the indexes of these publications.

WEEKLY

Aviation Week & Space Technology, a McGraw-Hill publication, 1221 Avenue of the Americas, New York, New York 10020. Perhaps the best source of information on the space industry.

Satellite Week, a newsletter from Television Digest, 1836 Jefferson Place N.W., Washington, D.C. 20036. News on direct broadcast satellite developments and new technology, coverage of international developments, details about domestic and international regulation (and deregulation), satellite marketplace intelligence, lists of mergers, acquisitions, and new business plans.

Space Calendar, a publication of Space Age Review, 3210 Scott Boulevard, Santa Clara, California 95051. A newsletter that discusses the major events related to space activities and provides information on such subjects as space entrepreneurs, activities at NASA facilities, space age technologies, and political action.

Important articles also appear in *Barron's, Business Week, Forbes, Newsweek,* and *Time.*

BIWEEKLY

Communications International, published by Dawson-Butwick Publishers, 1001 Connecticut Avenue N.W., Suite 301, Washington, D.C. 20036. Reports on international communications policy.

Space Business News, 1401 Wilson Boulevard, Suite 910 Arlington, Virginia 22209. From Pasha Publications, an independent newsletter and computer services firm. Reports on who is investing in space, their strategies, and the new products and services being developed.

Space Commerce Bulletin, from Television Digest, 1836 Jefferson Place N.W., Washington, D.C. 20036. Coverage on national and international space agencies, space business finance and insurance, space transportation and payloads, budget, contract and RFP monitor, federal government activity, space industrialization, materials processing, and remote sensing, among others.

MONTHLY

DBS NEWS, published by Phillips Publishing, 7315 Wisconsin Avenue, Suite 1220N, Bethesda, Maryland 20814. Covers DBS and related satellite broadcasting news. Phillips also publishes a more general newsletter on satellite management, marketing, and regulation, called *Satellite News;* an annual *Satellite Directory* with the names, addresses, telephone numbers, and key facts about the satellite industry; and *Space Enterprise Today,* which reports on technological progress, economic growth, industrial expansion, and profits in space commerce.

General monthly periodicals such as *Omni, Science Digest,* and *Space World,* also carry articles on space commerce and exploration.

ORGANIZATIONAL RELEASES

Foreign

European Space Agency publishes a monthly bulletin and various press releases. Information on launch services is available from the U.S. offices at 955 L'Enfant Plaza S.W., Suite 1404, Washington, D.C. 20005.

International Telecommunication Union, Documentation Section, Public Relations Division, General Secretariat, Place des Nations, Ch–1211, Geneva 20, Switzerland, puts out *Teleclippings,* a selection of articles on telecommunications from the press around the world.

Space Policy, an international quarterly journal published in February, May, August, and November, by Butterworth Scientific Ltd., P.O. Box 63, Westbury House, Bury Street, Guildford, Surrey GU2 5BH, United Kingdom. International articles on the space industry.

U.S. Government

U.S. Congress generates hearings and reports. Committees to watch are the House Committee on Science and Technology, and the Senate Committee on Commerce, Science, and Transportation.

Congressional Research Service, Library of Congress, Washington, D.C. 20540, publishes excellent reports on space activities and related issues as requested by Congress.

NASA, Washington, D.C. 20546, has various information services; and *Spinoff,* an annual report on space applications and technology in the private sector.

GROUP PUBLICATIONS

There is a growing number of professional organizations whose proceedings, panels, seminars, workshops, symposia, and colloquia provide the latest thinking in their respective fields. For example, the American Institute on Aeronautics and Astronautics (AIAA), 1663 Broadway, New York, New York 10019, publishes *Aerospace America.* This monthly magazine presents periodic reports on commercial space applications, the NASA budget, and decisions affecting both. AIAA also sponsors meetings in the field, which are announced in the magazine. Similarly, the American Association for the Advancement of Science (AAAS), 1515 Massachusetts Ave. N.W., Washington, D.C. 20005, puts out the weekly *Science*, with frequent references to space applications and activities in Washington.

Other relevant professional organizations are the American Astronautical Society, American Society of International Law, the American Bar Association Aerospace Law Committee, and the American Bar Association Section on Science and Technology. Added to this list are the publications of groups organized by citizens interested in expanding manned activities in space. They include the following:

- *L–5 News,* monthly magazine of the L–5 Society, 1060 East Elm, Tucson, Arizona 95719.
- *Insight,* bimonthly publication of the National Space Institute, West Wing Suite 203, 600 Maryland Avenue S.W., Washington, D.C. 20024.
- Space Studies Institute, 195 Sciences Building, Cornell University, Ithaca, New York 14853.
- Planetary Society, 302 Space Sciences Building, Cornell University, Ithaca, New York 14853, Carl Sagan, director.
- Center for Space Policy, 1972 Massachusetts Avenue, Cambridge, Massachusetts 02140. Publishes the monthly *Space Investment Report* on emerging investment opportunities in space commerce and executive summaries of its research in the space field.

SPECIAL PROCEEDINGS

Conferences each year generate proceedings containing articles on the latest technology, applications, economics, politics, law, and policy. An example is the annual conference of the International Astronautical Federation, whose proceedings are published annually by the AIAA. Other excellent proceedings are:

- Satellite Summit Conference, an annual event sponsored by *Satellite Week,* featuring presentations on the strategic engineering, marketing, and general business plans of satellite communications and space business.
- Financing Business in Space Conference, March 26–27, 1984, sponsored by the Center for Space Policy and Phillips Publishing. The *Executive Digest,* featuring industry leaders and highlights, is available from Phillips Publishing.
- Space Enterprise Videoconference, June 7, 1984, sponsored by BizNet, the television systems of the U.S. Chamber of Commerce, 1511 K Street N.W., Washington, D.C. 20005. Discussions of space commercialization issues. Papers available from BizNet.

DATA BASES

Many computerized data bases (e.g., Dialog, Infotrac) also have rich information on space commercialization. Satnet International Inc., 1010 Vermont Avenue, N.W., Suite 710, Washington, D.C. 20005, plans to offer data base services. Examples are Washington Space on legislative initiatives and regulatory activities concerned with the commercial potential of space technology and SatScan on satellite plans, capacity, and market projections. Other publications and organizations mentioned in Appendix A are planning their own data services.

Appendix B

Firms Involved in Space Commerce

Over 400 firms—aerospace as well as nonaerospace—are involved in space commerce. I have assembled a list (not exhaustive) of the major ones, compiled from numerous government documents and many other sources, such as *Aviation Week & Space Technology, Aerospace America, Aeronautics and Astronautics, Technology Review, Science Digest, Space World, The Futurist, Business Week, Time, Forbes, The Wall Street Journal,* the *Economist,* and other publications mentioned in Appendix A. The addresses, telephone numbers, and other useful information about the following firms can be found in the following sources:

Directory of American Firms Operating in Foreign Countries. Uniworld Business Publications, Inc., 50 East 42nd Street, New York, N.Y. 10017

Million Dollar Directory: America's Leading Public & Private Companies. Dun's Marketing Services, Inc., a company of the Dun & Bradstreet Corporation, Three Century Drive, Parsippany, N.J. 07054

Standard & Poor's Register of Corporations, Directors and Executives. Standard & Poor's Corporation, 25 Broadway, New York, N.Y. 10004

Ward's Business Directory. *U.S. Private Companies. Largest Private Plus Selected Public Companies.* Information Access Co., 11 Davis Drive, Belmont, Calif. 94002

ANALYSIS AND PROMOTION SERVICES

Aerospace Corp.

American Institute of Aeronautics and Astronautics

Andrews & Kurth

Arthur D. Little
Astro Research
Battelle Columbus Labs
Booz-Allen Hamilton
Center for Space Policy
Chase Econometrics
Communications Center of Clarksburg
Compucon
Computer Sciences
Comsearch
Coopers-Lybrand
Delta Vee/The Space Co.
ECON
Ecosystems International
Edmund Young
Encom System
Euroconsult
Frost & Sullivan
Future Systems
Management and Technical Services
Mathematica
Miller Communications (Canada)
Mitre
National Chamber Foundation
Nixon Professional Services
Planet Earth Productions
Planning Research
Racal, Decca (UK)
RAND
R&D Associates
Robert Shaw Associates
Rockwell International
Science Applications (SAI)
Space Contracts Institute
Spacewatch
SRI International
Stanford Research Institute
Stern Telecommunications
Stockdale and Associates
Systems Analysis

Taglar and Associates
Teleconsult International
Universities Space Research Associations
U.S. universities near NASA locations
Utah Innovation Center
Weinberg Consulting Group
Wyle Labs

DEFENSE CONTRACTORS AND SUBCONTRACTORS

U.S. Operators

Alpha Industries
EDO
Frequency Electronics
Geltech
General Defense
General Dynamics
General Electric
Grumman Aerospace
Helionetics
Hughes Aircraft
IBM
McDonnell Douglas
Optelecom
RCA
Rockwell International
Tech-Sym
TRW
Westinghouse

Non-U.S. Operators

British Aerospace
Marconi Ltd. (UK)

EARTH OBSERVATIONS

U.S. Operators

Aeros Data Corp.
American Science & Technology

BDM Corp.
Bendix Field Engineering
COMSAT
Earth Satellite
Fairchild Space & Electronics
General Electric
Geosat Committee
Hughes Communications
Integral Systems
IBM
Lockheed Missiles & Space
Miltope Corp.
RCA Astro-Electronics Division
Space Access Corp.
Space America
Sparx
Terra-Mar

Non-U.S. Operators

Dornier System GmbH (FRG)
SPOT Image (France)
Swedish Space

FINANCING

Aetna Diversified Technologies
A. G. Becker Paribus
Bank of America
Bankers Trust
Brentwood Associates
Chase Manhattan
Citicorp
First Boston
First Chicago
Greylock Management
Kleinwork, Benson
Lehman Bros. Kuhn, Loeb
Merrill Lynch–White Weld
Morgan Stanley
Northwest Association of Minneapolis

J. R. Packer
Private Export Financing
Prudential-Bache Securities
Rothschild
Saloman Bros.
Satellite Financial Planning
Sconset Group
Shearson-American Express
Stenbeck Reassurance (Sweden)
Wertheim & Co.

INSURANCE

Alexander and Alexander
Barring Aviation (UK)
Chadbourne, Parke, Whiteside & Wolfe
Crowley Warren, Co. (UK)
Faugere & Jutheau (France)
Frank B. Hall, Inc.
Gras Savoye S.A. (France)
IBM
Inspace (Corroon & Black)
International Technology Underwriters
Johnson & Higgins
Lloyd's of London (UK)
Marsh-McLennan
Sedgwick Group (UK)
U.S. Aviation Underwriters
Willis, Faber & Dumas, Ltd. (UK)

MICROGRAVITY MANUFACTURING R&D

Allegheny International
Aluminum Co. of America (ALCOA)
Battelle Columbus Labs
Beckman Instruments
Bethlehem Steel
W. S. Brown
Calcitet
Celanese
Du Pont

Eastman Kodak
Ecosystems International
EG&G
Eli Lilly & Co.
Gellman Research Associates
General Motors
Grumman Aerospace
GTI
Honeywell
International Nickel
International Space
John Deere & Co.
Johnson & Johnson (Ortho Pharmaceutical Division)
Johnson Mathey
Kaiser
KMS Fusion
Lovelace Medical Foundation
Marvaland
McDonnell Douglas Astronautics
Microgravity Research Associates
Monsanto
Northrop Aircraft
Parker-Hannifin
Rockwell International Space Div.
G. D. Searle & Co.
Space Studies
A. E. Staley Manufacturing
Terramar
3M Corp.
TRW
Union Carbide
United States Steel
Wang
Westech Systems
Westinghouse

PAYLOAD AND OTHER SERVICES

Architects Equities
Astrospace
Astrotech International

Ball Aerospace
Boeing Aerospace Comcon Systems
Cometto Industriale SPA (Italy)
Device Engineering
Eagle Engineering
Fairchild Space & Electronics
Getaway Special Services
Grumman Aerospace
GTI
Hughes Aircraft
Instrument Technology Associates
International Space
Orbital Systems
Pacific American Satellite
Prospace (France)
Rockwell International
Satellite Systems Engineering
Spaceco Ltd.
Space Industries Inc.
Space Projects
Space Services International
Teledyne Brown
Wyle Laboratories

SATELLITE COMMUNICATIONS

U.S. Satellite Manufacturers

Fairchild Space & Electronics
Ford Aerospace & Communications
General Electric Space Division
Globesat, Inc.
Hughes Aircraft Space & Communications Group
Lockheed Missiles & Space
RCA Astro-Electronics Division
Rockwell International Space Division
TRW
3M Corporation

Non-U.S. Satellite Manufacturers

Aeritalia (Italy)
Aerospatiale (France)

Allgemeine Electricitats Gesellschaft
ANT Nachrichtentechnik GmbH (FRG)
Bell Telephone Mfg. Co. (Belgium)
British Aerospace Dynamics Group (UK)
Communications Research Center (Canada)
Compagnia Nazionale Satellite di Telecommunicazioni (Italy)
Compagnie Industrielle Radioelectrique (Switzerland)
Dornier System GmbH (FRG)
ERNO Raumfahrttechnik GmbH (FRG)
Fabrica Italiana Apparecchiature Radioelettrische (Italy)
Fokker (Netherlands)
Kampsax International (Denmark)
Laben (Italy)
Marconi Space & Defense Systems (UK)
Matra SA (France)
Messerschmitt-Boelkow-Blohm (FRG)
Microtecnica (Italy)
Mitsubishi Electric (Japan)
NEC (Japan)
Nippon Electric (Japan)
Selenia Spazio (Italy)
SENER (Technica Industrial/Naval S.A., Spain)
Sharp Corp. (Japan)
Siemens AG (FRG)
Société Anonyme Belge de Constructions (Belgium)
Société Nationale Industrielle Aerospace (France)
Spar Aerospace (Canada)
Standard Elektrik Lorenz (FRG)
Telefunken (FRG)
Thomson CSF (France)
Toshiba Corp. (Japan)

Ground-Station Equipment Manufacturers

ANT Nachrichtentechnik GmbH (FRG)
Antenna Development & Manufacturing
ATC
Avcom
BEI Electronics
Byers Communications Systems
ComDev Ltd.

Comtech Antenna
Comtech Data
R. L. Drake
Electromagnetic Sciences
Encom Systems
General Instrument of Canada
GTE International Systems
Harris
Hoosier Electronics
International Phasor Telecom (Canada)
LNR Communications
Local Digital Distribution
MacDonald Dettricher & Associates (Canada)
Magnatech
Megasat (UK)
Microdyne
Microwave Services International
Modulation Associates
Motorola
NHK Tokyo
Norsat International (Canada)
Oak Communications
Philips Public Telecommunications
Potomac Marine and Aviation
Satcom Technologies
Satellite Telecommunications Systems (Italy)
Satellite Transmission & Reception Specialists
Scientific Atlanta
Stanford Telecommunications
TCI
Thomson CSF (France)
Varian
Video Electronics
Vitalink
Wegener Communications

U.S. Satellite Communication Services

Advanced Business Communications
Alascom
American Telephone & Telegraph

Argo Communications
Bonneville Satellite Communications
Catholic Telecommunications Network of America
Columbia Communications
COMSAT
Connex International
DBS
Dataspeed
Digital Telesat
EMX Telecom
Equatorial Communications
Federal Express
Focus Broadcast Satellite
Ford Aerospace Satellite Service
Graphic Scanning
GTE Satellite
GTE Spacenet
HBO Network
Hughes Communications Galaxy
International Satellite
Kellogg Communications
M/A-Com Research Center
Mitel
Modern Telecommunications
National Christian Network
National Exchange
Netcom
Oak Media
Orion Satellite
Private Satellite Network
Public Service Consortium
Rainbow Satellite
RCA Americom
Satellease
Satellite Business Systems
Satellite Development Trust
Satellite Syndicated Systems
Satellite Television
Satellite Television PLC
Seatel

Services by Satellite
Skyband
Southern Pacific Communications
Southern Satellite Systems
Space Communications
Systematics General
Teleconcepts in Communications
Telemedia International
Teleport Communications
Telstar
Turner Communications
United Satellite Communications
United States Satellite Broadcasting
United States Satellite Systems
Videonet
Video Satellite Systems
Videostar Connections
Western Union & Telegraph
World Communications
X Ten (IBM)

Non-U.S. Satellite Communication Services

Brightstar Communications (UK)
Canadian Satellite Communications
Detecon (FRG)
Eurospace
Eutelsat
GTS (UK)
International Maritime Satellite Communications Organization
International Satellite Telecommunications Organization
Japan Telegraph & Telephone
Mercury Communications (UK)
National Post & Telecommunications Industry (PRC)
Satritec (Ivory Coast)
Satel Conseil (France)
Swedish Space
Telecommunications Satellite Corp. of Japan
Telekable Vienna (Austria)
Telesat Canada
Telespazio (Italy)

United Satellites (UK)
Visnews (UK)

Satellite Navigation Services

Dataspeed
Geostar
Holiday Inn Hinet
Intelmet
Martin Marietta Communications Systems
Mobile Satellite
National Satellite Paging
Stanford Telecommunications
Texas Instruments

SPACE TRANSPORTATION

U.S. Operators

Aerojet Tech Systems/Ford Aerospace & Communications
American Rocket Company
Boeing Aerospace
Commercial Cargo Spacelines
Computer Sciences
Cyprus
Earth-Space Transport Systems
Fedex Spacetran
General Dynamics
Grumman Aerospace
Lockheed Space Operations
Martin Marietta
McDonnell Douglas Astronautics
Morton-Thiokol
Orbital Sciences
Orbital Systems
Pacific American Launch Services
Phoenix Engineering
Rockwell International Space Division
Satellite Propulsion
Space Projects
Space Services Inc. of America

Space Transport
Space Vector
Starstruck
Stiennon Partners
Transpace
Transpace Carriers
Truax Engineering
United Space Boosters
United Technologies Chemical Systems Division

Non-U.S. Operators

Aeritalia Divizione Spazio (Italy)
Arianespace (France)
BPD Difesa Spazio (Italy)
Messerschmitt-Boelkow-Blohm (UK)
Mitsubishi Heavy Industries (Japan)
OTRAG (FRG)

SPIN-OFFS

American Science and Engineering
Analytichem International
Celanese Corporation
Composite Consultation Concepts Inc.
Fisher Pen Company
Intellinet Corporation
Owens-Corning Fiberglass Corporation
Scott Aviation
Structural Composites Industries
3M Corporation
Wyle Laboratories

Glossary

Ablation The melting away of a spacecraft's heat shield during reentry.

Aerobrake An air brake used to slow a spaceship with the upper layers of a planet's atmosphere to conserve the spaceship's propellants.

Aerospace Includes the atmosphere and the regions of space beyond it.

Aphelion The point farthest from the Sun in the path of a solar satellite.

Apocynthion The point farthest from the Moon in the orbit of a lunar satellite.

Apogee The point farthest from Earth in the orbit of an earth satellite.

Artificial intelligence (AI) The discipline of developing and applying computer systems to produce characteristics usually associated with intelligent behavior (e.g., understanding language, learning from experience, logical reasoning, problem solving, and explaining its own behavior).

Artificial satellite A spacecraft that circles the Earth or other celestial body. The term is usually shortened to *satellite*, but it then also applies to natural moons.

Astro A prefix meaning "star." It also means "space" in such words as *astronautics* (the science of space flight).

Astronaut A U.S. space pilot.

Attitude The position of a spacecraft in relation to some other point, such as the spacecraft's direction of flight, the position of the Sun, or the position of the Earth.

Automation The use of electronic or mechanical machines to perform routine functions with minimal human intervention.

Base A permanently occupied center for people on the Moon, Mars, or in space that provides life support and work facilities; bases would evolve from outposts. *See* Outpost.

Biosatellite An artificial satellite that carries animals or plants.

Biosphere The total environment of Earth that supports self-sustaining and self-regulating human, plant, and animal life or an artificial closed-ecology system in which biological systems provide mutual support and recycling of air, water, and food.

Booster The propulsion system that provides most of the energy for a spacecraft to go into orbit.

Brownian movement The irregular motion of a body arising from the thermal motion of the molecules of the material in which the body is immersed. Such a body will of course suffer many collisions with the molecules, which will impart energy and momentum to it. Because there will be fluctuations in the magnitude and direction of the average momentum transferred, the motion of the body will appear irregular and erratic. In principle, this motion exists for any foreign body suspended in gases, liquids, or solids.

Burnout The point in the flight of a rocket when its propellant is used up.

Capsule A manned spacecraft or a small package of instruments carried by a larger spacecraft.

Carbonaceous A type of meteorite or asteroid containing significant percentages of water, carbon and nitrogen—essential elements that when processed would permit humanity to increase its independence from Earth.

Closed-ecology life support system (CELSS) A mechanical or biological system that recycles the air, water, and food needed to sustain human life on a space station or base.

Cosmonaut A Soviet space pilot.

Cycling spaceship A space station designed for human habitation that permanently cycles back and forth between the orbits of Earth and Mars.

Eccentricity The variation of a satellite's path from a perfect circle.

Ecosystem A community of humans, plants, and animals together with their physical environment.

Escape velocity The speed a spacecraft must reach to coast away from the pull of gravity.

Exhaust velocity The speed at which the burning gasses leave a rocket.

Galaxy An irregular, elliptic, disk- or spiral-shaped system containing billions of stars. Earth is situated in a spiral-shaped galaxy called the Milky Way, one of billions of galaxies in the universe.

Gantry A special crane or movable tower used to service launch vehicles.

Geostationary earth orbit A circular orbit approximately 22,300 miles above Earth's surface in the plane of the equator. An object in such an orbit rotates at the same rate as the planet and therefore appears to be stationary with regard to any point on Earth's surface. It is a specific type of geosynchronous orbit.

Heat shield A covering on a spacecraft to protect the craft and astronaut from high temperatures of reentry.

Heliosphere The large region of space influenced by the Sun's solar wind and the interplanetary magnetic field. This vast sea of electrical plasma may extend as far as 10 billion miles from the Sun, affecting the magnetospheres, ionospheres, and upper atmospheres of Earth and other Solar System bodies.

Hypergol Propellants that ignite when mixed together.

Inner Solar System The part of the Solar System between the Sun and the main asteroid belt. It includes the planets Mercury, Venus, Earth, and Mars—distinct from the outer planets, Jupiter, Saturn, Uranus, Neptune, and Pluto.

Libration points Unique points in space, influenced by gravitational forces of neighboring bodies, in which objects with the correct initial location and velocity remain fixed without significant expenditure of propellant. Also called Lagrange points, after the French mathematician who calculated their existence.

LOX or liquid oxygen A common oxidizer made by cooling oxygen to −183°C (−297°F).

Magnetosphere A region surrounding a planet, extending out thousands of miles and dominated by the planet's magnetic field so that charged particles are trapped in it.

Mass driver An electromagnetic accelerating device for propelling solid or liquid material, for example, from Earth's Moon into space or for providing propulsion by ejecting raw lunar soil or asteroidal material as reaction mass.

Microgravity An extremely low level of gravity. As experienced by shuttle crews, for example, one-millionth the level of gravity on Earth's surface.

Module A single section of a spacecraft that can be disconnected and separated from other sections.

Orbit The path of a satellite in relation to the object around which it revolves.

Orbital maneuvering vehicle A device used much like a harbor tug in ship operations, with remotely controlled manipulator arms to handle spacecraft and refueling operations with great care.

Outpost An initial location to provide shelter for a few people on the Moon or Mars; it may not be permanently occupied.

Oxidizer A substance that mixes with the fuel in a rocket, furnishing oxygen that permits the fuel to burn.

Pericynthion The point closest to the Moon in the orbit of a lunar satellite.

Perigee The point closest to Earth in the orbit of an Earth satellite.

Perihelion The point closest to the Sun in the path of a solar satellite.

Period The time it takes for a satellite to make one revolution.

Propellant A substance burned in a rocket to produce thrust. Propellants include fuels and oxidizers.

Reentry That part of a flight when a returning spacecraft begins to descend through the atmosphere.

Rendezvous A space maneuver in which two or more spacecraft meet.

Revolution One complete cycle of a heavenly body or an artificial satellite in its orbit.

Robotics The use of automated machines to replace human effort, although they may not perform functions in a human-like manner.

SCRAMJET A supersonic combustion ramjet engine that can operate in the hypersonic region of flight (ten times the speed of sound).

Settlement A permanent community of humans in space or on the surface of the Moon or Mars with life support, living quarters, and work facilities; it will evolve from a base. *See* Base.

Spacecraft An artificially created object that travels through space.

Spaceport A transportation center in space that acts like an airport on Earth. It provides a transport node where passengers or cargo can switch from one spaceship to another and a facility where spaceships can be berthed, serviced, and repaired.

Specific impulse A measurement of engine performance. It is the ratio of the pounds of thrust produced by the engine, minus the drag from the engine, per pounds of fuel flowing through the engine each second.

Stage One of two or more rockets combined to form a launch vehicle.

Tele-operator A system equipped with its own propulsion system, television camera, and equipment that can be remotely operated. *See* Orbital maneuvering vehicle.

Telepresence The use of real-time video communications coupled with remote control techniques that would provide an operator on Earth's surface or other location with the capability to carry out complex operations in space or on the surface of a planet or Moon.

Telescience Conducting scientific operations in remote locations by teleoperation.

Thrust The push given to a rocket by its engine.

Unpiloted A spacecraft without human operators.

Van Allen radiation The high-energy charged particles trapped by the geomagnetic field that form belts of intense radiation in space about the Earth. The belts consist primarily of electrons and protons and extend from a few hundred kilometers above the Earth to a distance of about 8 R_e (R_e = radius of Earth = 6371 km = 3959 mi). The belts of trapped particles were discovered by James Van Allen and his coworkers in 1958 through radiation detectors carried on satellites Explorer 1 and 3. Many additional experiments have shown that the radiation has a complicated, time-dependent structure.

References

Adler, Lee (1966), "Symbiotic Marketing," *Harvard Business Review* 44 (November-December), 59–71.

Aerospace America (1986), "Space Insurance: Who Takes the Risk?" (March), 12–14, 16–17.

Allen, Joseph (1984), *Entering Space: An Astronaut's Odyssey*. New York: Stewart, Tabori & Chang.

Altman, Robert (1984), "The Food and Drug Administration: Regulation in Space," *Food Drug Cosmetic Law Journal* (October), 445–60.

American Cancer Society, New York, N.Y.

American Institute for Aeronautics and Astronautics (1977), *Space: A Resource for Earth*. New York: American Institute for Aeronautics and Astronautics.

Anderson, John W. (1986), "Soviet Strides: American Launchers are Grounded, But the USSR Program Couldn't Be Healthier," *Commercial Space* (Summer), 26–31.

———. (1987), "Competition in the Communications Market Exposes the Payoffs, Problems and Perils of Direct Broadcast," *Commercial Space* (Winter), 53–57.

Apollo Expeditions to the Moon (1975). NASA publication. Washington D.C.: Government Printing Office.

Argyris, Christopher (1962), *Interpersonal Competence and Organizational Effectiveness*. Homewood, Ill.: Richard D. Irwin.

Astronomy (1985), "Stopping Space Sickness" (February), 60.

Aviation Week & Space Technology (1984a), "Council Sets Japan's Space Goals" (March 12), 130.

———. (1984b), "Growth Trends: U.S. Aerospace Industry" (March 12), 16–17.

———. (1984c), "Chinese Reserve Ariane Payload Space" (April 30), 18.

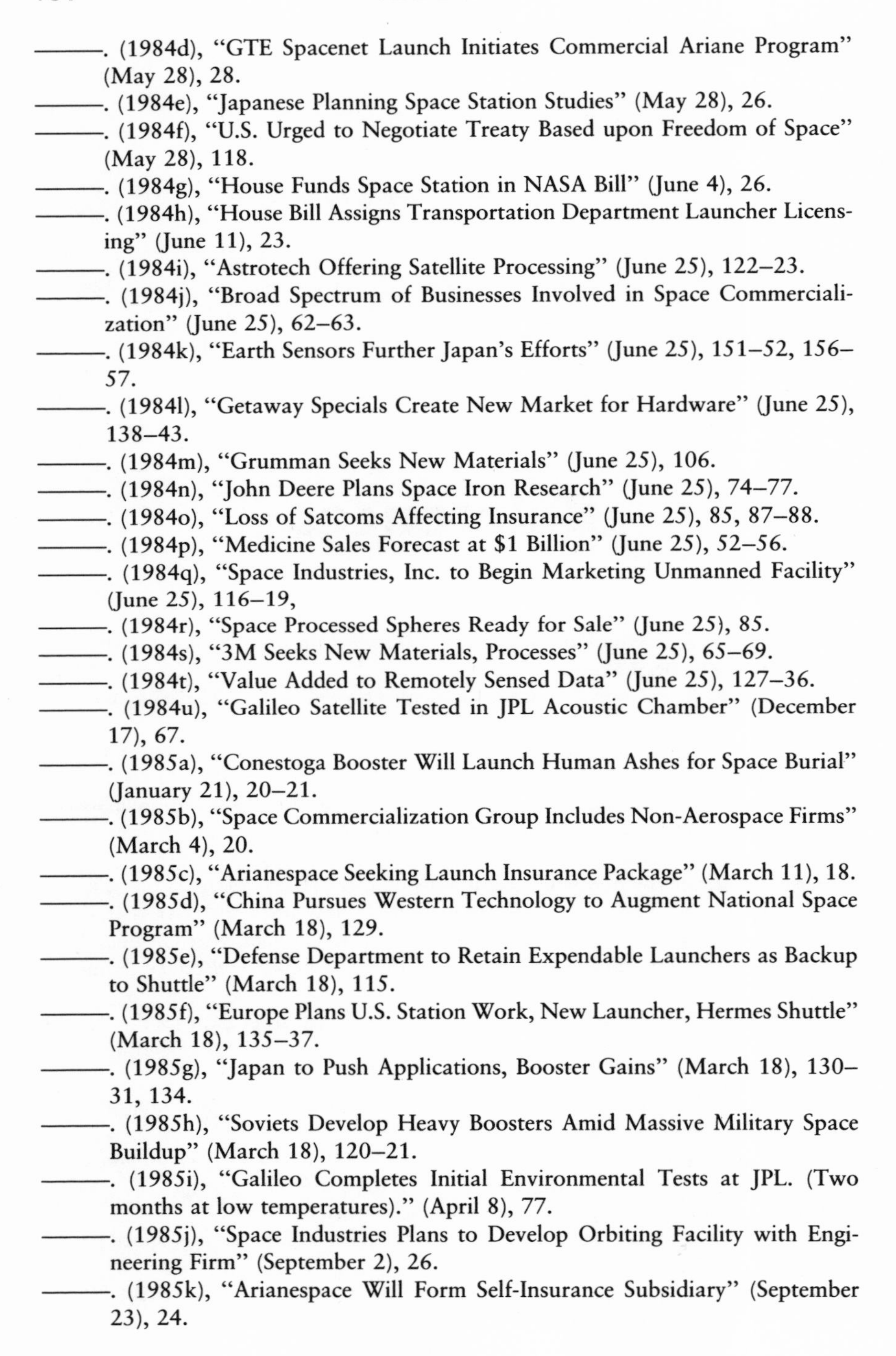

———. (1984d), "GTE Spacenet Launch Initiates Commercial Ariane Program" (May 28), 28.
———. (1984e), "Japanese Planning Space Station Studies" (May 28), 26.
———. (1984f), "U.S. Urged to Negotiate Treaty Based upon Freedom of Space" (May 28), 118.
———. (1984g), "House Funds Space Station in NASA Bill" (June 4), 26.
———. (1984h), "House Bill Assigns Transportation Department Launcher Licensing" (June 11), 23.
———. (1984i), "Astrotech Offering Satellite Processing" (June 25), 122–23.
———. (1984j), "Broad Spectrum of Businesses Involved in Space Commercialization" (June 25), 62–63.
———. (1984k), "Earth Sensors Further Japan's Efforts" (June 25), 151–52, 156–57.
———. (1984l), "Getaway Specials Create New Market for Hardware" (June 25), 138–43.
———. (1984m), "Grumman Seeks New Materials" (June 25), 106.
———. (1984n), "John Deere Plans Space Iron Research" (June 25), 74–77.
———. (1984o), "Loss of Satcoms Affecting Insurance" (June 25), 85, 87–88.
———. (1984p), "Medicine Sales Forecast at $1 Billion" (June 25), 52–56.
———. (1984q), "Space Industries, Inc. to Begin Marketing Unmanned Facility" (June 25), 116–19,
———. (1984r), "Space Processed Spheres Ready for Sale" (June 25), 85.
———. (1984s), "3M Seeks New Materials, Processes" (June 25), 65–69.
———. (1984t), "Value Added to Remotely Sensed Data" (June 25), 127–36.
———. (1984u), "Galileo Satellite Tested in JPL Acoustic Chamber" (December 17), 67.
———. (1985a), "Conestoga Booster Will Launch Human Ashes for Space Burial" (January 21), 20–21.
———. (1985b), "Space Commercialization Group Includes Non-Aerospace Firms" (March 4), 20.
———. (1985c), "Arianespace Seeking Launch Insurance Package" (March 11), 18.
———. (1985d), "China Pursues Western Technology to Augment National Space Program" (March 18), 129.
———. (1985e), "Defense Department to Retain Expendable Launchers as Backup to Shuttle" (March 18), 115.
———. (1985f), "Europe Plans U.S. Station Work, New Launcher, Hermes Shuttle" (March 18), 135–37.
———. (1985g), "Japan to Push Applications, Booster Gains" (March 18), 130–31, 134.
———. (1985h), "Soviets Develop Heavy Boosters Amid Massive Military Space Buildup" (March 18), 120–21.
———. (1985i), "Galileo Completes Initial Environmental Tests at JPL. (Two months at low temperatures)." (April 8), 77.
———. (1985j), "Space Industries Plans to Develop Orbiting Facility with Engineering Firm" (September 2), 26.
———. (1985k), "Arianespace Will Form Self-Insurance Subsidiary" (September 23), 24.

———. (1985l), "Underwriter Ends Its Space Launch Coverage" (September 23), 24.

———. (1985m), "Companies Reassess Space-Based Drug Processing Amid Bioengineering Advances" (September 30), 50, 54.

———. (1985n), "Intec Offers Launch-Phase Insurance" (September 30), 28.

———. (1985o), "Launcher Company, Travel Agency Reach Space Pact" (September 30), 24.

———. (1985p), "Satellite Losses May Cause Reinsurance Capacity Drop" (September 30), 50.

———. (1985q), "EOSAT Develops Marketing Rules for Landsat Data" (November 4), 48, 53.

———. (1985r), "Insurance Manager Sees Further Aviation, Space Policy Limits" (November 4), 67.

———. (1985s), "Manufacturers Form Organization to Protect Proprietary Rights" (November 4), 67.

———. (1985t), "Soviet Mission to Mars" (December 9), 16.

———. (1986a), "French Guiana Space Center Facility Increases Ariane Launch Capability" (March 31), 126–27.

———. (1986b), "Soviets Expected to Launch Space Shuttle by 1987" (March 31), 26.

———. (1988a), "Commerce Report Cites Potential of Commercial Space Markets" (June 6), 17.

———. (1988b), "Thiokol Drops Out of Bidding For Advanced Shuttle Solid Rocket" (June 13), 19.

———. (1988c), "Plessey, Westland Win SDI Contracts To Study Advanced Radar, Thermoplastics" (June 20), 21.

———. (1988d), "Government Payloads Dominate Commercial Launch Manifest" (July 4), 24.

———. (1988e), "NASA Studies Unmanned Space Shuttle Missions" (July 11), 27.

———. (1988f), "Strategic Defense Facility To Open at Sandia in 1989" (August 22), 81.

———. (1988g), "Soviets to Deploy U.S.-Built Amsat Satellite From Mir in 1989" (August 29), 19.

Ayres, Robert (1969), *Technological Forecasting and Long-Range Planning*. New York: McGraw-Hill.

Bagozzi, Richard P. (1975), "Marketing As Exchange," *Journal of Marketing* 39 (October), 32–39.

———. (1986), *Principles of Marketing Management*. Chicago: Science Research Associates.

Baier, Martin (1983), *Elements of Direct Marketing*. New York: McGraw-Hill.

Banks, Howard (1988), "It's Time to Bust Up NASA," *Forbes* (February 8), 101–8.

Bauer, Raymond A., and Dan H. Fenn, Jr. (1972), *The Corporate Social Audit*. New York: Russell Sage Foundation.

Beauchamp, Marc (1987), "The U.S. Hasn't Fired an Interplanetary Space Shot in a Decade. So Why Is NASA's Jet Propulsion Laboratory So Busy? Down to Earth," *Forbes* (June 15), 74, 78.

Bekey, Ivan (1983), "Tethers Open New Space Options," *Aeronautics and Astronautics* 21 (April), 22–40.

———. (1984), "Applications of Space Tethers." New York: American Institute of Astronautics.

———, and Daniel Herman, eds. (1985), *Space Stations and Space Platforms—Concepts, Design, Infrastructure, and Uses.* New York: American Institute of Aeronautics and Astronautics.

Bennett, Peter D. (1988), *Marketing.* New York: McGraw-Hill.

Beyond the Atmosphere: Early Years of Space Science (1980). NASA publication. Washington, D.C.: Government Printing Office.

Bierer, Loretta Kett (1987), "Researchers Are Studying How Our Bodies React to Long Stays in a Weightless Environment." *Commercial Space* (Winter), 46–49.

Bitto, Ron (1981), "Cosmic Counselor," *Omni Magazine* (August), 48–51, 94.

Blau, Thomas, and Daniel Gouré (1984), "Military Uses and Implications of Space," *Society* 21 (January-February), 13–17.

Bock, Gordon (1987a), "Run Silent, Run to Moscow: Congress Protests the Sale of High-tech Secrets to the Soviets," *Time* (June 29), 45.

———. (1987b), "Limping Along in Robot Land: A Once Hopeful U.S. Industry Goes Awry," *Time* (July 13), 46–48.

Boston, Penelope, ed. (1984), *The Case for Mars.* San Diego, Calif.: American Astronautical Society.

Bowen, Howard R. (1953), *Social Responsibilities of the Businessman.* New York: Harper.

Browning, Michael (1986), "Space Race Winner—by Default," *Miami Herald* (October 2), 13A.

Bullock, Chris (1984) "Big Business in Space? NASA Moves to Smooth the Path of New Comers," *Interavia* (April), 389.

———, and Paul W. Rubin (1984), "Satellite Telecommunications: The Ground Business Grows," *Interavia* (November), 1232.

Bundy, McGeorge; George F. Kennan; Robert S. McNamara; and Gerard Smith (1984), "The President's Choice: Star Wars or Arms Control," *Foreign Affairs* 63, no. 2 (Winter), 264–78.

Business Week (1981), "A Bargain-Basement Challenge to NASA," (June 22), 42.

———. (1987a) "A Data Base for Star Wars Technology" (November 9), 106.

———. (1987b), "Starship Enterprise Chasing NASA's Unfinished Business" (November 9), 98–99.

———. (1988), "The White House Sets Its Sights on Mars" (February 8), 34.

Byars, Carlos (1985a), "Rockwell Wins NASA Shuttle Contract," *Houston Chronicle* (September 13), Section 3, p. 1.

———. (1985b), "Astronauts to Build Structures in Space during Shuttle Flight," *Houston Chronicle* (October 27), Section 1, p. 2.

———. (1985c), "Shuttle with Crew of 8 Successfully Launched into Orbit," *Houston Chronicle* (October 30), Section 1, p. 10.

———. (1985d), "Crew Puts Spacelab Tests Back on Line," *Houston Chronicle* (November 1), Section 1, p. 10.

———. (1985e), "Space-made Drug Said to Be Kidney Hormone," *Houston Chronicle* (November 19), Section 1, p. 7.

———. (1985f), "Atlantis Astronauts Launch 1st Satellite," *Houston Chronicle* (November 27), Section 1, p. 2.

———. (1986a), "Challenger Explodes; Shuttle Falls into Ocean; Crew Apparently Killed," *Houston Chronicle* (January 28), Section 1, pp. 1, 4.

———. (1986b), "Investigators Find Solid Clues Pointing to Cause of Blast," *Houston Chronicle* (February 1), Section 1, p. 25.

Carreau, Mark and William E. Clayton, Jr. (1986), "Veteran Astronaut Hits Space Station Planning," *Houston Chronicle* (July 19), Section 1, p. 16.

———, and Bill Dawson (1986), "Astronauts Re-examine Role in Space Program," *Houston Chronicle* (April 6), Section 1, pp. 1, 5.

Center for Space Policy (1985), "Commercial Space Industry in the Year 2000: A Market Forecast." Unpublished paper.

Chandler, David (1987), "Discoveries Alter Understanding of Sun: Wrinkles, Bright Spots Reflected in New Theory," *Miami Herald* (February 18), p. 2E.

Chase, Victor D. (1986), "Physicians' Arsenal," *Commercial Space* (Summer), 36.

Christol, Carl Q. (1980), "International Liability for Damage Caused by Space Objects," *American Journal of International Law* (Fall), 346–71.

———. (1982), *The Modern International Law of Outer Space*. New York: Pergamon Press.

———. (1984), "Space Law: Justice for the New Frontier," *Sky & Telescope* (November), 406–9.

Churchill, Gilbert A. (1987), *Marketing Research: Methodological Foundations*. New York: Dryden Press.

Clark, Evert, and Seth Payne (1987), "America's Heavyweights Are Flying Higher Than Ever—For Now," *Business Week* (November 9), 99.

Clarke, Arthur (1968), *2001: A Space Odyssey*. New York: New American Library.

Clausen, Peter, and Michael Brower (1987), "The Confused Course of SDI," *Technology Review* (October), 60–72.

Cohen, Dorothy (1986), "Trademark Strategy," *Journal of Marketing* 50 (January), 61–74.

Cohen, Maxwell (1964), *Law and Politics in Space*. Montreal: McGill University Press.

Coleman, Herbert J. (1985a), "Satellite Makers Predict Recovery in Insurance Market, High Premiums," *Aviation Week & Space Technology* (November 11), 197, 199–201.

———. (1985b), "Insurance," *Commercial Space* (Fall), 61.

Collins, Michael (1985), "An Apollo 11 Astronaut Addresses the Question of Man vs. Machine," *Commercial Space* (Summer), 67–72.

Commercial Space (1986), "R&D Update. Spacelab DA Results" (Summer), 75.

———. (1987a), "Drug Makers Press Plans for Future Space Processing" (Winter), 58–59.

———. (1987b), "Europeans Are Putting the Final Touches on a Broad, Multibillion-Dollar Space Package Deal" (Winter), 27–29.

———. (1987c), "The Station Is Raising Lots of Questions about Space Law" (Winter), 43–45.

———. (1987d), "3M's Ten-Year Agreement with NASA Provides for Major Research in Orbital Manufacturing" (Winter), 50–52.

Cooper, Jr., Henry S. F. (1976), *A House in Space*. New York: Holt, Rinehart & Winston.

Coopersmith, Jonathan (1988), "Space: The Russian Frontier," *Wall Street Journal* (April 25), Section 2, p. 26.

Corddry, Charles W. (1985), "Technological Leap: Star Wars Research Could Bring Enormous Benefits to Medicine," *Houston Chronicle* (October 6), Section 1, p. 8.

Covault, Craig (1984a), "China Offers Space Launch Services," *Aviation Week & Space Technology* (October 8), 48, 52, 57.

———. (1984b), "Unique Products, New Technology Spawn Space Business," *Aviation Week & Space Technology* (June 25), 40–41, 44–45, 47, 49, 51.

———. (1985a), "Burgeoning Space Launch Capability is Leading toward Economic Competition," *Commercial Space* (Fall), 18–21.

———. (1985b), "Shuttle/Station Cost Challenges Key to Future Space Operations," *Aviation Week & Space Technology* (March 18), 109–11, 113.

———. (1986), "Japan: Emerging Space Power," *Commercial Space* (Summer), 16–22.

———. (1987a), "International Cooperation in Space," *Commercial Space* (Winter), 16–19.

———. (1987b), "Moon Base Gaining Support as New U.S. Space Goal," *Aviation Week & Space Technology* (May 11), 22–24.

———. (1988), "Record Soviet Manned Space Flight Raises Human Endurance Questions. 326 days," *Aviation Week & Space Technology* (January 4), 25.

Cravens, David W. (1987), *Strategic Marketing*. Homewood, Ill.: Richard D. Irwin.

Crown, Judith (1985), "NASA to Consolidate Shuttle Services in One Contract," *Houston Chronicle* (October 25), Section 3, pp. 1, 3.

Cunningham, Ann Marie (1987), "Business versus Star Wars," *Technology Review* (May–June), 17.

Dall'Acqua, Joyce (1988), "The Search for Extraterrestrial Intelligence," *Futurist* (May-June), 16–18.

Dalrymple, Douglas J., and Leonard J. Parsons (1986), *Marketing Management: Strategy and Cases*. New York: John Wiley & Sons.

David, Leonard (1982), "International Moon Base Proposed," *Space World* (May), 12–13.

———. (1986), "The Next 50 Years Will Bring about Massive Changes in Uses of Space," *Commercial Space* (Fall), 36–39.

Davis, Neil W. (1983), "Japan in Space," *Space World* (May), 9.

Dawson, Bill (1986), "Military Space Chief Lauds Manned Flights," *Houston Chronicle* (March 29), Section 1, p. 16.

DeGeorge, Gail (1986), "1,108 Losing Jobs at Space Center," *Miami Herald* (September 5), 1E.

Deudney, Daniel (1982), "Space Industrialization: The Mirage of Abundance," *Futurist* (December), 47–53.

Diamond, Edwin (1964), *The Rise and Fall of the Space Age*. Garden City, N.Y.: Doubleday.

Dickson, David (1983), "After Spacelab, Europe wants a Better Deal," *Science* 222 (December 9), 1099–1100.

Doig, Stephen K. (1987a), "NASA Plans 50-year Mission to Measure Distances between Stars," *Miami Herald,* (January 8), 17A.

———. (1987b), "Disaster's Legacy Is Pervasive," *Miami Herald* (January 29), 1A, 10A.

Dooling, Dave (1982), "The First Space Factories," *Space World* (March), 4–7, 32–33.

———. (1985a), "Orbital Processing," *Commercial Space* (Summer), 14–20.

———. (1985b), "Orbital Processing Promises Investors Immediate Financial Rewards," *Commercial Space* (Summer), 14–20.

———. (1986), "Materials Processing," *Commercial Space* (Summer), 66–70.

Dorr, Jr., Les (1985), "Satellite Insurance in the 80's," *Space World* (February), 28–30.

———. (1987), "Russki Business," *Space World* (July), 31–34.

Downey, Arthur, J., ed. (1985), *The Emerging Role of the U.S. Army in Space.* Washington, D.C.: National Defense University Press.

Drucker, Peter F. (1954), *The Practice of Management.* New York: Harper & Row.

Dula, Art (1985), "Private Sector Activities in Outer Space," *International Lawyer* (Winter), 159–87.

Dye, Lee (1985), "Miracle in Space: Troublesome Satellite Finally Finds Its Orbit," *Houston Chronicle* (November 3), Section 1, p. 2.

Eason, Henry (1984a), "How a Space Venture Could Ease Suffering on Earth," *Nation's Business* (June), 48.

———. (1984b), "The Challenge of Business in Space," *Nation's Business* (June), 46–47.

Eberhart, Jonathan (1985), "Japan Launches Probe to Comet Halley," *Science News* (January 12), 22.

Economist (1984a), "Factories in Space?" (August 4), 16.

———. (1984b), "NASA Looks for Ways of Boosting Business into Orbit" (August 4), 73, 76.

———. (1984c), "Where Apples Fall Upwards" (August 4), 74.

———. (1985), "Malfunction" (April 27), 97.

———. (1987a), "Science and Technology: Per Ardua and ESA" (November 7), 93.

———. (1987b), "The Great Siberia in the Sky" (October 3), 93–94, 96.

Egan, John J. (1985), "What's the Payoff? Pluses and Minuses of Space Processing," *Commercial Space* (Summer), 62–66.

———. (1986), "How Has the Shuttle Challenger Accident Affected the Business? In Stock Prices These Have Been Trying Times," *Commercial Space* (Fall), 32–34.

Engel, J. F.; H. F. Fiorillo; and M. A. Cayley (1972), *Market Segmentation: Concepts and Applications.* New York: Holt, Rinehart and Winston.

Enis, Ben M., and Keith K. Cox (1985), *Marketing Classics: A Selection of Influential Articles.* Boston: Allyn and Bacon.

Ennals, Richard (1987), *STAR WARS: A Question of Initiative.* New York: John Wiley & Sons.

Etzioni, Amitai (1964), *The Moon-Doggle: Domestic and International Implications of the Space Race.* Garden City, N.Y.: Doubleday.

Far Travelers: The Exploring Machines (1985). NASA publication. Washington, D.C.: Government Printing Office.

Feazel, Michael (1984a), "Europe Pushes Space Station Role," *Aviation Week & Space Technology* (June 18), 16–17.

———. (1984b), "Sparx Decision Clouds Imaging Projects," *Aviation Week & Space Technology* (June 25), 147–49.

———. (1985a), "Germany Approves Participation in Space Station, New Ariane," *Aviation Week & Space Technology* (January 21), 17–19.

———. (1985b), "German Minister Proposes Initiative to Improve European Defenses, *Aviation Week & Space Technology* (December 9), 19–20.

Financial World (1984), "New Frontiers: Looking for Profits in Space" (November 14), 25.

Finch, Jr., Edward Ridley, and Amanda Lee Moore (1985), *ASTROBUSINESS: A Guide to the Commerce and Law of Outer Space*. New York: Praeger Publishers.

Finegan, Jay (1987), "STAR WARS, INC.," *Inc.* (April), 68–76.

Fink, Donald (1985), "Crossroads for Commercialization," *Aviation Week & Space Technology* (November 25), 9.

Foley, Theresa M. (1986a), "Family of Challenger Pilot Files $15 million Claim against NASA," *Aviation Week & Space Technology* (July 21), 29–30.

———. (1986b), "U.S. Manufacturers Begin the Job of Rebuilding the U.S. Space Program: ELVs" *Commercial Space* (Fall), 16–21.

———. (1987a), "Canadians Making Early Effort to Organize for Station Use," *Aviation Week & Space Technology* (May 11), 24–25.

———. (1987b), "Space Station Partners Are Trying to Work Out Agreements Covering the Next 10 Years," *Commercial Space* (Winter), 41–42.

Frank, R. E.; W. F. Massey; and Y. Wind (1972), *Market Segmentation*. Englewood Cliffs, N.J.: Prentice-Hall.

Frazer, Lance (1986), "Yours, Mine or Ours: Who Owns the Moon?" *Space World* (November), 24–26.

Froehlich, Walter (1985), *Space Station: The Next Logical Step*. NASA EP–213. Washington, D.C.: NASA.

Frost, Kenneth J., and Frank B. McDonald (1984), "Space Research in the Era of the Space Station," *Science,* 226 (4681), 1381–85.

Futurist (1984), "Plans for U.S. Space Station" (April), 73–74.

———. (1985), "Space: The Long-Range Future" (February), 36–38.

———. (1988a), "Japanese Experts Predict" (May-June), 35.

———. (1988b), "Next Steps in Space" (May-June), 19–22.

———. (1988c), "The Aero-Space Plane" (May-June), 47.

Galloway, Eilene (1982), "The History and Development of Space Law." Report at Unispace '82, Vienna, Austria (August).

Ganoe, William H. (1987), "Industrial Space: The Modeler," *Space World* (July), 35.

Geneen, Harold, with Alvin Moscow. (1984), *Managing*. New York: Avon Books.

Gengtao, Chen (1987), "China's Great Wall Industrial Corporation Is Marketing Launch Services Worldwide," *Commercial Space* (Winter), 30–31.

Gersteenfeld, Arthur, ed. (1979), *Technological Innovation: Government/Industry Cooperation*. New York: Wiley.

Goldman, Marshall I. (1987), "The Shifting Balance of Power," *Technology Review* (April), 20–21.

Goldsmith, Donald, ed. (1980), *The Quest for Extraterrestrial Life—A Book of Readings*. Mill Valley, Calif.: University Science Books.

Goodrich, Jonathan N. (1988), "Privatization in America," *Business Horizons* 31 (January-February), 11–17.

Goodrich, Jonathan N.; Robert L. Gildea; and Kevin Cavanaugh (1979), "A Place for Public Relations in the Marketing Mix," *MSU Business Topics* (Autumn), 53–57.

Goodrich, Jonathan N.; Gary H. Kitmacher; and Sharad R. Amtey (1987), "Business in Space: The New Frontier?" *Business Horizons* 30 (January-February), 75–84.

Govoni, Stephen J. (1986a), "SDI's Momentum," *Financial World* (March 4), 28.

———. (1986b), "The Race for Profits in Space," *Financial World* (March 4), 20–28.

Grace, J. P. (1984), *War on Waste: President's Private Sector Survey on Cost Control*. New York: Macmillan.

Greeley, Jr., Brendan M. (1985), EOSAT Develops Marketing Rules for Landsat Data," *Aviation Week & Space Technology* (November 4), 48, 53.

Green, Paul E. (1978), *Analyzing Multivariate Data*. Hinsdale, Ill.: Dryden Press.

———, and Donald S. Tull (1978), *Research for Marketing Decisions*. 4th ed. Englewood Cliffs, N.J.: Prentice-Hall.

Greene, C. Scott, and Paul Miesing (1984), "Public Policy, Technology, and Ethics: Marketing decisions for NASA's Space Shuttle," *Journal of Marketing* 48 (Summer), 56–67.

Gregory, William H. (1986), "Researchers at MIT Say It's a Matter of Ease," *Commercial Space* (Summer), 58–60.

Grey, Jerry (1979), *Enterprise*. New York: Morrow.

Grissom, Virgil "Gus" (1968), *Gemini: A Personal Account of Man's Venture into Space*. New York: Macmillan.

Gupte, Pranay (1987), "Russia: Arms Merchant to the World," *Forbes* (November 2), 168, 172, 174.

Hair, Jr., Joseph F.; Ralph E. Anderson; and Ronald L. Tatham (1987), *Multivariate Data Analysis, with Readings*. New York: Macmillan.

Haley, Andrew G. (1963), *Space Law and Government*. New York: Appleton-Century-Crofts.

Hall, R. C. (1969), "Rescue and Return of Astronauts on Earth and in Outer Space," *American Journal of International Law* (Spring), 197–210.

Hanke, Steve H. (1984), *The Private Provision of Public Services and Infrastructure*. Report submitted to U.S. AID. Baltimore: Johns Hopkins University.

Harbison, Frederick, and Charles Myers (1959), *Management in the Industrial World*. New York: McGraw-Hill.

Harrison, Allen F., and Robert M. Bramson (1982), *Styles of Thinking*. Garden City, N.Y.: Anchor Press/Doubleday.

Harvard Business Review: On Management (1975). New York: Harper & Row.

Healey, Robert (1987), "Star Wars Makes Much More Sense as a Bargaining Chip Than as Shield," *Miami Herald* (October 31), 17A.

Herzberg, Frederick (1968), *Work and The Nature of Man*. Cleveland: World.

Higher Education Advocate (1987), "Star Wars: Faculty Push to Change Foreign Policy" (April 17), 1.

Houston Chronicle (1985a), " 'Railroad into Space' on Drawing Board" (November 12), Section 1, p. 10.

———. (1985b), "Space Commercialization Seen as Blossoming New Industry" (December 5), Section 1, p. 18.

———. (1985c), "Britain Becomes First Ally to Participate in Star Wars" (December 7), Section 1, p. 24.

———. (1986a), "Europeans Launch Two Satellites" (February 23), Section 1, p. 24.

———. (1986b), "NASA Says O-Rings Never Fully Tested in Cold" (March 6), Section 1, p. 2.

———. (1986c), "China Signs Deal to Launch Swedish Satellite" (March 29), Section 1, p. 22.

———. (1986d), "17th Ariane Rocket Launched after 3 Delays" (March 29), Section 1, p. 22.

Huszagh, Sandra, and Hiram C. Barksdale (1986), "International Barter and Counter Trade: An Exploratory Study," *Journal of The Academy of Marketing Science* 14 (Spring), 21–28.

Hyman, William A. (1966), *Magna Carta of Space*. Amherst, Wis.: Amherst Press.

Jain, Pravin C. (1983), *Strategic/Tactical Communications by Satellites*. Falls Church, Va.: Satellite Systems & Technology.

James, Barrie G. (1985), *Business Wargames*. Middlesex, England: Penguin Books.

Jasentuliyana, N., and R. Lee (1979), *Manual on Space Law*. New York: Oceana Publications.

Jastrow, Robert (1984) "Why We Need a Manned Space Station," *Science Digest* (May), 41–42, 92, 94.

Jenks, Wilfred C. (1965), *Space Law*. New York: Praeger.

Johnson, Diane (1986), "University of Colorado Research Center Is Creating Techniques to Help People Understand What Happens If Technical Processes Are Performed in a State of Microgravity," *Commercial Space* (Fall), 68–69.

Karas, Thomas (1983), *The High Ground*. New York: Simon and Schuster.

Katz, Daniel, and Robert Kahn (1966), *The Social Psychology of Organizations*. New York: John Wiley & Sons.

Kelly, Thomas (1987), "Science or Subsidy?" *Technology Review* (February-March), 10–11.

Kitmacher, Gary H. (1985), "Space Commercialization: Its Development." Working Paper (May). Houston: NASA/Johnson Space Center.

Kolcum, Edward H. (1984a), "Company Plans to Manufacture Crystals in Space," *Aviation Week & Space Technology* (June 25), 100–101.

———. (1984b), "NASA Defining Technology Required by Space," *Aviation Week & Space Technology* (May 14), 80.

———. (1985), "Space Station Manager's Next Big Job Is to Drum Up Business," *Commercial Space* (Summer), 81–85.

Koltz, Charles (1983), "Mining the Moon," *Space World* (February), 30–31.

Kotler, Philip (1971), *Marketing Decision Making: A Model-Building Approach*. New York: Holt, Rinehart and Winston.

———. (1984), *Marketing Management: Analysis, Planning and Control*. 5th ed. Englewood Cliffs, N.J.: Prentice-Hall.

———. (1988), *Marketing Management: Analysis, Planning, Implementation, and Control*. Englewood Cliffs, N.J.: Prentice-Hall.

———, Liam Fahey, and S. Jatusripitak (1985), *The New Competition.* Englewood Cliffs, N.J.: Prentice-Hall.

Kozicharow, Eugene (1984a), "Air Force Development Plan on Expandable Launchers," *Aviation Week & Space Technology* (June 18), 70, 74, 79.

———. (1984b), "Patent Law Finds Space Application," *Aviation Week & Space Technology* (June 25), 97–98.

Kuzela, Lad (1984), "Who'll Win the Race for Profits in Space?" *Industry Week* (August 6), 28–31.

Large, Arlen J. (1982a), "Space Shuttle's Flight Rates Will Soar in 1985 in a Bid to Cover Operating Costs," *Wall Street Journal* (June 16), 14.

———. (1982b), "Small Projects Get a Chance on the Shuttle," *Wall Street Journal* (June 24), 31.

Lehmann, Donald R., and Russell S. Winer (1988), *Analysis For Marketing Planning.* Plano, Texas: Business Publications.

Lemonick, Michael D. (1987a), "Supernova! Scientists Are Agog over the Brightest Exploding Star in 383 Years," *Time* (March 2), 60–69.

———. (1987b), "The Soviets Blast Out in Front. Energia's Launch Widens Moscow's Edge in Rocketry," *Time* (June 1), 58.

———. (1988), "Can They Escape Next Time?" *Time* (February 29), 68.

Lenorovitz, Jeffrey M. (1984) "Europe Outlines Future Space Plan," *Aviation Week & Space Technology* (June 25), 16–17.

———. (1985a), "Europe's Ariane Is Moving the International Community to Launch Autonomy," *Commercial Space* (Fall), 22–25.

———. (1985b), "French Space Agency to Select Hermes Manned Shuttle Contractor by Midyear," *Aviation Week & Space Technology* (March 11), 19–20.

———. (1986), "Soviets Test Spectrometric Laser System," *Aviation Week & Space Technology* (June 16), 92–93.

———. (1987), "Advances by the Soviet Union in Space Cooperation and Commercial Marketing Made 1986 a Landmark Year," *Commercial Space* (Winter), 20–22.

Lerner, Eric (1986), "Space Insurance—Who Takes the Risk?" *Aerospace America* (March), 12–17.

Levitt, I. M. (1959), *Target for Tomorrow: Space Travel of the Future.* New York: Fleet Publishing Corporation.

Levitt, Theodore (1983), "After the Sale Is Over," *Harvard Business Review* 83 (September-October), 87–93.

Lewis, Paul (1984), "Europe's 'Factories in Space,'" *New York Times* (May 14), p. D10.

Ley, Willie (1969), *Events in Space.* New York: David McKay Co.

Lilien, Gary L., and Philip Kotler (1983), *Marketing Decision Making: A Model-Building Approach.* New York: Harper & Row.

Living and Working in Space: A History of Skylab. (1983). NASA publication. Washington, D.C.: Government Printing Office.

Lovelock, Christopher H. (1984), *Services Marketing: Text, Cases and Readings.* Englewood Cliffs, N.J.: Prentice-Hall.

Lowndes, Jay C. (1985), "U.S. Opposes Rigid Planning of Geostationary Orbit, Spectrum," *Aviation Week & Space Technology* (November 4), 69, 71–73.

———. (1987), "Mixing Astronauts from Many Nations by the U.S. on Space Shuttle Missions Is Resulting in a New Version of the Melting Pot," *Commercial Space* (Winter), 34–35, 38–40.

Manber, Jeffrey (1986), "Crisis in Satellite Insurance," *STV* (March), 81–83.

Mann, Paul (1984), "Beggs Foresees Shuttle in Private Hands," *Aviation Week & Space Technology* (June 25), 20–21.

———. (1985), "Washington Broadens Its Efforts to Aid Small Business," *Commercial Space* (Summer), 21, 24–25.

Marcom, John Jr. (1984), "Satellite Company's Technology Is Too Far Ahead of Its Time," *Wall Street Journal* (May 3), 31.

Mark, Hans (1986), "Warfare in Space." *America Plans for Space.* Washington, D.C.: National Defense University Press.

Marketing News (1985), "AMA Board Approves New Marketing Definition" (March 1), 1.

Marketing Staff of the Ohio State University (1965), "A Statement of Marketing Philosophy," *Journal of Marketing* (January), 43.

Marsh, Alton K. (1984), "Space Services Pushing Conestoga Launch Vehicle," *Aviation Week & Space Technology,* (June 25), 163, 165.

Marshall, Tyler (1987), "Deaths of 3 Scientists Raise British Eyebrows," *Miami Herald* (April 8), 1A, 7A.

Martian Landscape (1978). NASA publication. Washington, D.C.: Government Printing Office.

McCarthy, E. Jerome, and William D. Perreault, Jr. (1987), *Basic Marketing: A Managerial Approach.* 9th ed. Homewood, Ill.: Richard D. Irwin.

McDougal, Myers, S.; Harold D. Lasswell; and Ivan A. Vlasic (1963), *Law and Public Order in Space.* New Haven, Conn.: Yale Press.

McDougall, Walter A. (1985), *The Heavens and the Earth: A Political History of the Space Age.* New York: Basic Books.

McGinley, Laurie (1987a), "NASA Picks Space-Station Main Builders," *Wall Street Journal* (December 2), 4.

———. (1987b), "The Latest Fashion for U.S. Spacemen Is Still Up in the Air," *Wall Street Journal* (November 2), 1, 22.

McKie, Robin (1985), "Life Beyond Earth?" *World Press Review* 32 (April), 54.

Mentzer, John T. and Forrest S. Carter (1985), *Readings in Marketing Today.* New York: Harcourt Brace Jovanovich.

Mereson, Amy (1984), "The Longest Arm of the Law: Nations Grapple with Interplanetary Legal Questions," *Science Digest* (September), 30.

Merrifield, John T. (1986), "Market Demand for Small Satellites Spurs Commercial Efforts at Globesat," *Commercial Space* (Summer), 61–63.

Meyer, David S. (1987), "French Spot and the U.S. Landsat Jockey for Position in the Race for a Multimillion-Dollar Remote Sensing Market," *Commercial Space* (Winter), 62–66.

Miami Herald (1986), "U.S. Awards Britain $8.7 Million in Star Wars Research Contracts" (December 9), 9A.

———. (1987a), "A Space First: Soviets Swap Orbiting Ship Crew", (December 21), 1A, 20A.

———. (1987b), "NASA Plans 50-year Mission to Measure Distances between Stars" (January 8), 17A.

———. (1987c), "Reagan OKs $10.9 Billion for Scaled-Down Space Station" (April 4), 3A.
———. (1987d), "Voyager 2 Shifts Course" (March 14), 7A.
———. (1988a), "New NASA Jet to Carry Shuttle" (March 1), 16A.
———. (1988b), "Report: Soviets Studying Push into Deep Space" (March 1), 16A.
Miglicco, Gary J. (1985), "Shuttle Launches of Satellites Are Making Space a Bottomline Business," *Commercial Space* (Summer), 36–39.
Minsky, Marvin, ed. (1985), *Robotics.* Garden City, N.Y.: Omni Press.
Mordoff, Keith H. (1984), "Shuttle Hitchhiker Spurs Business Effort," *Aviation Week & Space Technology* (June 25), 102–5.
Morrison, Philip, ed. (1973), *The Search for Extraterrestrial Intelligence.* NASA SP–419. Washington, D.C.: Government Printing Office.
Morrow, Jr., Walter E. (1987), "SDI Research Is Critical," *Technology Review* (July), 24–25, 77.
NASA (1976), *A Forecast of Space Technology 1980–2000.* NASA SP–387. Washington, D.C.: NASA (January).
NASA Tech Briefs. New York: Associated Business Publications.
National Center for Health Services, Hyattsville, Maryland.
Nelson, Bill (1987), "The Space Station Involves the American Visit as Well as a Technological Imperative," *Miami Herald* (October 4), 1C, 4C.
North, David M. (1984), "Brazil Plan to Launch Its Own Satellites in the 1990's," *Aviation Week & Space Technology* (July 9), 60–61.
Nozette, Stewart, and Robert Lawrence Kahn, eds. (1987), *Commercializing SDI Technologies.* New York: Praeger Publishers.
O'Lone, Richard G. (1984), "Bay Area Firms Pursue Booster Designs," *Aviation Week & Space Technology* (June 25), 166–169.
———. (1985), "Entrepreneurial Spirit Combines with Hard-Headed Business Sense," *Commercial Space* (Fall), 37–39.
O'Neill, Gerard K. (1976), *The High Frontier: Human Colonies in Space.* New York: Morrow.
———. (1981), *2081: A Hopeful View of the Future.* New York: Simon & Schuster.
Orwell, George (1949), *1984.* New York: Harcourt Brace Jovanovich.
Osborne, David (1985), "Business in Space," *Atlantic Monthly* (May), 45–53, 56–58.
Parasuram, A. (1986), *Marketing Research.* Reading, Mass.: Addison-Wesley.
Payne, Seth (1985), "Space Insurance Premiums Are Headed for Orbit," *Business Week* (September 30), 53.
———. (1987), "The Space Station Launching a Thousand Contracts," *Business Week* (November 23), 126–27, 130.
Pesavento, Peter (1987), "Sputnik's Heirs: What the Soviets Are Doing in Space," *Technology Review* (October), 26–35.
Peterson, Robert A. (1988), *Marketing Research.* Plano, Texas: Business Publications.
Peterson, Thane (1987), "Europe's Aerospace Companies Are in Seventh Heaven," *Business Week* (November 23), 54.
Pioneering The Space Frontier (1986). National Commission on Space. New York: Bantam Books.
Poole, Lynn (1958), *Your Trip into Space.* New York: McGraw-Hill.

Poole, Jr., R. C. (1980), *Cutting Back City Hall.* New York: Universe Books.

Preiss, Byron, ed. (1985), *The Planets.* New York: Bantam Books.

Preston, Lee E., ed. (1968), *Social Issues in Marketing.* Oakland, N.J.: Scott, Foresman.

———. ed. (1980), *Business Environment Public Policy 1979 Conference Papers.* St. Louis: AACSB.

Pride, William M., and O. C. Ferrell (1987), *Marketing: Basic Concepts and Decisions.* 5th ed. Hopewell, N.J.: Houghton Mifflin.

Rapport, Daniel (1981), "Who Owns Space?" *Space World* (August-September), 4.

Reichhardt, Tony (1982), "Why We Explore the Planets, *Space World* (January), 26–28.

Reynolds, Glenn Harlan (1987), Review of *Space Stations and the Law: Selected Legal Issues—Background Paper. Jurimetrics Journal* (Summer), 431–42.

———., and Robert P. Merges (1985), "The Role of Commercial Development in Preventing War in Outer Space," *Jurimetrics: Journal of Law, Science and Technology* 25 (Winter), 130–46.

Ris, Howard (1987), "Faulty Marketing of Star Wars: White House Uses Fear to Sell Uncertain System," *Miami Herald* (September 1), 11A.

Roland, Alex (1987), "Manned Space Spectaculars Have Impoverished the Rest of Our Space Program," *Miami Herald* (October 4), 1C, 5C.

Roth, Gabriel (1985), *Private Provision of Public Services in Developing of Public Services in Developing Countries.* Washington, D.C.: World Bank Economic Development Institute.

Ryan, Cornelius, ed., (1952), *Across the Space Frontier.* New York: Viking Press.

Sagan, Carl (1980), *Cosmos.* New York: Random House.

Schwartz, Michiel, and Paul Stares, eds. (1985), *The Exploitation of Space.* London: Butterworth & Co.

Science Year 1986. Chicago: World Book.

Selling to NASA (1986). NASA publication. Washington, D.C.: Government Printing Office.

Shapiro, Stacy (1985a), "Satellite Owners Launch New Proposals to Cope with Rate, Capacity Crisis," *Business Insurance* (February 4), 3–4.

———. (1985b), "Birth of Satellite Coverage Was in the Cards: Barrett," *Business Insurance* (August 19), 3, 24.

———. (1985c), "Successful Repair of Disabled Probe Might Ease Rates," *Business Insurance* (August 19), 3, 24.

———. (1986), "Insurers May See Few Claims from Space Shuttle Disaster," *Business Insurance* (February 3), 1, 30.

Shifrin, Carole A. (1984), "Investors Taking Cautious View of Private Programs," *Aviation Week & Space Technology* (June 25), 78–80, 83.

Shulman, Seth (1987), "Biological Research and Military Funding," *Technology Review* (April), 13–14.

Shutak, Shannon (1986), "Willing and Able: Expendable Vehicle Manufacturers Gear for Anticipated Launch Demand," *Commercial Space* (Fall), 22–25.

Simon, Herbert (1957), *Administrative Behavior.* New York: Macmillan.

Simpson, Theodore, ed. (1985), *The Space Station: An Idea Whose Time Has Come.* New York: IEEE Press.

Simsarian, James (1963), "Outer Space Cooperation in the United Nations," *American Journal of International Law* (Winter), 854–67.

Skylab, Our First Space Station (1977). NASA publication. Washington, D.C.: Government Printing Office.

Smith, Bruce A. (1985a), "Galileo Undergoes Final Tests Prior to Shipment to Kennedy," *Aviation Week & Space Technology* (October 21), 135.

———. (1985b), "Hughes, Insurers Reach Accord on Leasat–3 Compensation Plan," *Aviation Week & Space Technology* (July 8), 24–25.

———. (1987), "New Launch Vehicle Could Help the U.S. Return to Space with Commercial Payloads," *Commercial Space* (Winter), 32.

Smith, David H. (1987), "Observing the Energetic Universe," *Technology Review* (May-June), 67–73.

Smith, Eleanor (1986), "Stress in Orbit (Psychological Care of Space-Station Crews)," *Omni* (November), 42–43.

Smith, W. R. (1956), "Product Differentiation and Market Segmentation as Alternative Marketing Strategies," *Journal of Marketing* 21 (July), 3–8.

Sobel, Dave (1987), "Astro Docs," *Omni* 9 (April), 22.

Space Commerce Bulletin (1985), "Space Insurance Industry Differs on Role of Government; Shortage Hurts Shuttle Most" (November 8), 5–6.

Space World (1983), "Spinoffs from the Space Shuttle" (June-July), 19–20.

———. (1987a), "Are the Profits Really There?" (July), 34.

———. (1987b), "Destination Mars: A Conversation with Michael Collins" (July), 16–20.

Spinoff 1986, NASA. Washington, D.C.: U.S. Government Printing Office.

Spitz, Stephen A. (1980), "Note," *Harvard International Law Journal* (Spring), 579–84.

Stanton, William J. (1984), *Fundamentals of Marketing*. 7th ed. New York: McGraw-Hill.

———., and Charles Futrell (1987), *Fundamentals of Marketing*. New York: McGraw-Hill.

Straubel, Michael S. (1987), "The Commercial Space Launch Act: The Regulation of Private Space Transportation," *Journal of Air Law and Commerce* 52 (Summer), 941–69.

Taylor, Diane (1987), "Ocean Sensing: Joint U.S./French Project May Aid in Weather Forecasting," *Commercial Space* (Winter), 60–61.

Taylor, Jr., L. B. (1987), *Commercialization of Space*. New York: Franklin Watts.

Technology Review (1987), "Japan in Space" (January), 11–12.

This Island Earth (1970). NASA publication. Washington, D.C.: Government Printing Office.

Thompson, Mark (1987), "Firm Accused of Faking Tests of Key Cruise Missile Parts," *Miami Herald* (December 22), 18A.

Time (1979), "The Sad State of Innovation" (October 22), 70–71.

———. (1982a), "Here Come the Europeans," (February 8), 63.

———. (1982b), "Outer-Space Entrepreneurs" (September 20), 19.

———. (1983), "Battlestar Columbia?" (April 17), 20.

———. (1984a), "Business Heads for Zero Gravity" (November 26), 30.

———. (1984b), "Competition in the Cosmos" (November 26), 31.

Trefil, James (1985), *Space Time Infinity*. Washington, D.C.: Smithsonian Books.

Udwadia, Firdaus E. (1987), "International Cooperation in Space: An Assessment of Economic Risks," *Technological Forecasting & Social Change* (April), 173–188.

U.S. Congress (1978). Senate. Committee on Commerce, Science, and Transportation. *Space Law: Selected Basic Documents*. 2d ed. Washington, D.C.: Government Printing Office.

———. (1980a). Senate. Committee on Commerce, Science, and Transportation. *Agreement Governing the Activities of States on the Moon and Other Celestial Bodies*. Washington, D.C.: Government Printing Office.

———. (1980b). Senate. Committee on Commerce, Science, and Transportation. *The Moon Treaty*. Washington, D.C.: Government Printing Office.

———. (1980c). House. Committee on Science and Technology. *The Space Industrialization Act of 1980*. Washington, D.C.: Government Printing Office.

———. (1982). House. Committee on Science and Technology. *National Space Policy*. Washington, D.C.: Government Printing Office.

———. (1983a). House. Committee on Science and Technology. *Review of Materials Processing in Space*. Washington, D.C.: Government Printing Office.

———. (1983b). House. Committee on Science and Technology. *Space Commercialization*. Washington, D.C.: Government Printing Office.

———. (1984a). Senate. Committee on Commerce, Science, and Transportation. *Commercial Space Launch Act*. Washington, D.C.: Government Printing Office.

———. (1984b). House. Committee on Science and Technology. *Initiatives to Promote Space Commercialization*. Washington, D.C.: Government Printing Office.

———. (1984c). House. Committee on Science and Technology. *International Cooperation and Competition in Space*. Washington, D.C.: Government Printing Office.

———. (1984d). House. Committee on Science and Technology. *The Expendable Launch Vehicle Commercialization Act*. Washington, D.C.: Government Printing Office.

———. (1984e). Office of Technology Assessment. *Civilian Space Stations and the U.S. Future in Space*. Washington, D.C.: Office of Technology Assessment (November).

———. (1985). Office of Technology Assessment. *International Cooperation and Competition in Civilian Space Activities*. Washington, D.C.: Office of Technology Assessment.

———. (1986). Office of Technology Assessment. *Space Stations and the Law: Selected Legal Issues*. Background Paper. Washington, D.C.: Government Printing Office.

U.S. Department of Defense (1983), *Soviet Military Power*. Washington, D.C.: Government Printing Office. [This report is published annually.]

U.S. News & World Report (1986), "Life in Space: It Won't Be the Ritz" (January 13), 72–73.

Varadarajan, P. "Rajan," and Daniel Rajaratnam (1986), "Symbiotic Marketing Revisited," *Journal of Marketing* 50 (January), 7–17.

Vassiliev, M. (1958), *Sputnik Into Space*. New York: Dial Press.

Vaucher, Marc E., and Kelly Robertson (1986), "The Space Shuttle Accident Forces Companies to Change Plans," *Commercial Space* (Fall), 42–49.

Wagner, Rice S. (1984), "The Lawmen Head for Space," *Space World* (May), 8–9.

Waldrop, M. Mitchell (1983), "The Commercialization of Space," *Science* 221 (September 30), 1353–54.

Wall Street Journal (1985), "U.S. Transfers Landsat to Venture of Hughes, RCA" (September 30), 4.

———. (1986a), "Insurance Rates on Satellites Seen Increasing" (January 29), 20.

———. (1986b), "Shuttle Accident Could Benefit Rival Space Programs in Europe" (January 29), 20.

———. (1986c), "Space" (January 29), 28.

———. (1986d), "Tragedy in Space: Explosion May Force a Rethinking of Need for Manned Mission" (January 29), 1.

———. (1986e) "Shuttle Contractors Could Face Suits from Explosion, But Liability Is Hazy" (January 30), 6.

———. (1986f), "Shuttle's Customers Are in Quandry over the Future of These Space Projects" (January 30), 6.

———. (1986g), "Shuttle Disaster Puts Spotlight on Safety" (February 4), 2.

Weaver, Lelland A. C. (1987), "Factories in Space: The Role of Robots," *Futurist* (May-June), 29–34.

Webb, David C. (1985), *Trends in the Commercialization of Space.* Washington D.C.: Aerospace Industries Association.

Webb, Tom (1987), "Aerospace Industry Profits Skyrocket," *Miami Herald* (December 17), 1E.

Welch, Joe L. (1980), *Marketing Law.* Tulsa, Okla.: Petroleum Publishing Co.

White, Jr., Harold M. (1984), "International Law and Relations." In T. Stephen Cheston, Charles M. Chafer, and Sallie Birket Chafer, eds., *Social Sciences and Space Exploration,* pp. 40–51, 101–3. Washington, D.C.: Georgetown University and NASA.

Williams, Sylvia M. (1987), "The Law of Outer Space and Natural Resources," *International and Comparative Law Quarterly* (January), 142–51.

Wolfe, Tom (1979), *The Right Stuff.* New York: McGraw-Hill.

Wyles, Randy (1987), "Fence Lines in Space," *Texas Business* (March), 54–55.

Zimmerman, Mark D. (1985), "Space Agenda to the Year 2000," *Machine Design* (January 10), 72–77.

Index

ABOUT THE AUTHOR

JONATHAN N. GOODRICH received his Ph.D. and M.B.A. degrees from the State University of New York at Buffalo, the M.A. from the University of Georgia, and the B.Sc. from the University of the West Indies, Jamaica. He has published extensively in journals such as *Harvard Business Review, Journal of Marketing, Journal of Marketing Research, Business Horizons, Business Topics,* the *Journal of Travel Research,* and *Proceedings* of marketing conferences in the United States and abroad. He is Associate Professor of Marketing at Florida International University, Miami. He taught previously at Indiana University and the University of Houston.